Small Power Planning: Blackout, Brownout, and Emergency Power Concepts

Small Power Solutions for when the power goes out and Off-Grid Living

Rodney Biddle

http://www.SmallPowerPlanning.com/

ISBN: 978-0-9996491-4-5

Imprint: Independently published

The web addresses cited in this book were current as of October 2020, unless otherwise noted.

Disclaimer:

The information given in this book is solely for educational and entertainment purposes and in no way should be considered to be legal, medical or professional advice. The author and publisher assume no liability for any actions taken by any individual or group that is considered illegal based on information in this book.

Revision: 2.5

This newest revision adds a section on performing a home energy audit, discussions on Rolling Blackouts and Rolling Blackout solutions, whole-home battery storage systems, and whole-home examples using the new Sol Donum Vulcan series of portable self-installable whole-home battery solutions.

Contents

Introduction

I am sitting here in the comfort of a local restaurant to write the updated introduction for the second revision of my newest book, *"Small Power Planning: Blackout, Brownout, and Emergency Power Concepts"*. This is an update to the Workshop book I published early this year after returning to the writing table and expanding the book to include additional sections focused on larger scale battery solutions capable of integrating directly into the home. I am reading stories about the expanding wildfires around Lake Tahoe and the damage from this past weekend's Hurricane Ida which hit New Orleans on the 16th anniversary of Hurricane Katrina. In Louisiana and surrounding areas, the predictions are already out that areas will be without power for weeks leaving residents to depend on emergency solutions, many of which are exactly what this book is about.

Challenges for about a million people this week and in the days and weeks to come for many of those will be locating fuel, keeping cool in expected high heat, keeping refrigeration working, keeping lighting on for security and protecting their homes and belongings. For many who have evacuated the challenges will also include just getting home and living in an infrastructure that has again been ravaged by nature requiring weeks or months to be fully restored again.

After writing and releasing the first edition of this book in early 2021 I attended an preparation planning conference up in the mountains of North Carolina. I had the pleasure at that conference of making a new friend who works with a new company producing expandable battery systems that can be used stand-alone or integrated into the home and connected to solar to provide long-term energy solutions ideal for someone living off-grid or in blackouts on a smaller scale than the huge power walls by companies such as Tesla or LG, but large enough to keep the basics up and running when needed. This was also at a time shortly after the February Texas freezes which knocked out power to millions of homes and created a rolling blackout situation like those also experienced this year in California and other western states due to high energy needs.

California – The state that is one of the heaviest pushers of electric and green vehicles, this year began asking residents to restrict their charging of their EV cars because their power companies could not meet energy demands. Washington DC is pushing for a movement to have up to 50% of vehicles on the road to be electric by 2030 – a huge goal because energy production continues to be a problem in many parts of the country with an aging energy infrastructure and an increasing demand for power as our population grows and expands in many areas of the country. In Florida I am witnessing the increase in large solar farms the size of neighborhoods cropping up in areas that just a few years ago were orange groves and agricultural areas, farms growing food now growing energy.

"Green" energy has its problems – Rainy and stormy days crashes energy production. High wind days in California forces the shutdown of wind farms – for days or even longer periods of time. California just this year has touted that they are up to 95% green now – but as they increase their level of green energy from sources such as solar and wind, what happens when those sources are unavailable?

In California on heigh wind days, the high winds often force the shutdown of the wind farms to prevent damage to wind turbines. When high winds get above about 55 MPH the windmills will be shut down to prevent damage. This is in a state where up to 36% of the energy now comes from Wind. Solar of course has its own problems that are well known due to rainy or cloudy days which cut energy production down to a small fraction. But have you heard about what is referred to as "The Photovoltaic Heat Island Effect"? This is where larger solar plants have a direct impact on local temperatures to the areas which are now raising global warming concerns.

The purpose of this book is to present solutions that are focused around planning a small power solution for short-term power during an event that creates a temporary power loss. In our discussions we will look at solutions that can provide partial power to your family for periods of up to 30 days. These discussions are not around long-term power solutions such as permanent windmills or whole-house solar, but rather smaller solutions that can be used for your home or taken as a portable solution in case the need to relocate to another location.

More often it is seeming that we are witnessing more problems that trigger power blackouts. From natural disasters such as our late summer hurricanes that can devastate areas for days to weeks, to rolling blackouts that seem to occur yearly now in California, to unforeseen events such as the 2021 Texas Freeze where the temperatures dropped so deep as to freeze natural gas lines and windmills cutting out power to millions in the affected areas. Without some level of power, we are not only left in the dark, but we are without sources of emergency heating and cooling often placing lives at risk. From refrigeration for preserving foods and critical medicines, to keeping the temperature warm or cool in at least a single room, or just being able to keep the news on or cell phones charged, having some level of power is both critical for not only our sanity but our safety.

In 2017 following hurricane Irma in Florida, power losses from the hurricane which traveled up the backbone of the state are considered to have caused over 400 deaths in nursing homes. After the hurricane passed through Florida pass a new state-wide requirement for all nursing homes to have emergency power capabilities permanently installed. Today as I write this, the death toll in Texas alone is still being calculated with over 80 deaths being attributed to the freezing temperatures, not counting the deaths in surrounding states also impacted by the extended deep freeze. What impacted Texas the most was not being ready for such cold temperatures that are common in areas farther north.

A new Normal?

One question we must ask – Is the risk of a power event a new normal? As time moves along our power grid continues to age. A 2015 report by the US Government concluded that 70% of power transformers are 25 years or older in age, and that 60% of circuit breakers are 30 years old or older. And what are we doing to update our grid? Current government direction is moving towards the shutdown of more fossil fuel driven electric generation, and the Green movement continues to shy away from clean Nuclear energy shifting more towards "Clean" Solar and wind production. But – Solar fails at night and bad weather days while wind fails on high-wind days or extreme cold-weather days. And neither are always great options depending on where you are.

But – That is the direction we are moving. I'm not saying that a green energy move is not positive but moving too quickly and shutting down our fossil fuel systems will place us at risk. That means that the grid, as well as having an aging problem, will be depending on less dependable green solutions in the future. What this also means is that having some level of self-reliance with power production and alternative energy sources will become more important in the future. In 2018 wind provided over 11.5% of California's electricity – 29% of the state's overall Renewables Portfolio Standard (RPS). As this number continues to grow, and instances occur which affect wind generation, there will naturally be a growing number of instances of rolling blackouts. In states such as California where wind is sufficient to be a source for alternative energy (Requiring about 6-7 miles per hour minimum), high winds more than 55 MPH can require the actual shut down of windmills to prevent them from literally flying apart. High-wind scenarios can also force the shutdown of power lines carrying electricity due to risk of fires from downed power lines.

Fuel Generators

A fuel-based generator can be set up for operation quickly – just run to the store, pick up the generator, fuel, oil and necessary power cords and you're operating. Most of the lower sized generators we will be looking at will operate on gasoline and/or propane, with some operating using diesel fuel. Strong advantages of a fuel generator include quick setup, and high levels of power limited mostly by availability of fuel. Portable generators can be picked up that can provide over 10-kW of power – enough to provide most of the power for your home and still be portable for relocation if necessary. A fuel-based generator will greatly outperform a renewable source such as solar or wind if you have fuel available.

The primary disadvantages of fuel generators include the need and dependency of a fuel source, and noise generation. The dependency on fuel means if there is a shortage of that fuel then your ability to rely on the generator fails. For this reason, many small generators are designed as "Dual Fuel" generators operating from either gasoline or propane fuel.

Noise produced by a loud generator is also often a concern – noise from a running generator can at times attract unwanted attention – including attention from criminals or anyone looking for a generator, fuel, or supplies. During hurricane IRMA reports of generators being stolen while even running at night were made, with one even including two fellows running down the road carrying a large generator only having to drop it when caught to escape being stopped.

Solar Generators – aka Power Stations/Power Boxes

A "Solar" Generator is more like a battery that can put out AC power for our normal home appliances than an actual generator. In fact – many companies are now referencing their units as "Power Stations", "Power Generators" or "Power Boxes". The Solar generator does not actually produce power itself, but rather contains a rechargeable battery than absorbs energy from an outside source. The source for power for a solar generator can be from Solar Panels, Windmills, or even from regular grid AC power charging. An advantage of the Solar generator is the lack of need of a traditional fuel source allowing it to be recharged and used thousands of times. Another advantage of this type of power source is silent operation and lack of dangerous fumes. Silent operation means the generator can be operated quietly at night or indoors. Lack of fumes from burning fuel means that the generator can be safely used indoors in close quarters or can be carried into an emergency shelter or even your bedroom for safe operation.

While the primary advantage is the lack of need for traditional fuel, the disadvantages come from the reliance on less reliable sources of energy. For solar – overcast or night hours result in a reduction of power production (Or no power at all being produced) and for windmills when the wind speed drops so does the energy generation. Cost is also a significant disadvantage – while a fuel generator can be purchased for just a few hundred dollars and up, a solar generator with the equivalent capacity will run much more for the generator and necessary solar panels. Another significant disadvantage is the amount of power that can be generated in a short amount of time. We will discuss all of these in depth when we get into our chapter on power planning.

A full home solar system designed to provide power to your home will be designed to produce much more electricity and for longer term and a more permanent, non-mobile solution. Though whole-home solar is a fantastic solution, it is a solution beyond the scope of this book and workshop, and therefore we will not be covering it in detail. We will touch on Whole-home solar however later in the Solar chapter of this book.

Goal and Layout of this book

This book is being written to be used as a part of an overall workshop focused on Emergency power planning and not for teaching long-term alternative power solutions. As an instructor covering general readiness planning which includes emergency communications, my goal is to provide short-term lifesaving levels of power – keep communications up and running including the ability to run a ham radio or just charge a cell phone – keeping lighting up for safety and security, and being able to keep some levels of heating, cooling, and refrigeration available.

Reference articles for more information:

1. Study: Hurricane Irma caused 400 senior deaths in Florida (ABC News)
https://abcnews.go.com/Health/wireStory/study-hurricane-irma-caused-400-senior-deaths-florida-73583102

2. Travis County medical examiner investigating at least 86 deaths in the wake of Texas freeze (Statesman.com)
https://www.statesman.com/story/news/2021/02/25/texas-power-outage-death-toll-medical-examiner-processing-86-cases/6808150002/

3. Aging grids drive $51B in annual utility distribution spending (UtilityDrive.com)
https://www.utilitydive.com/news/aging-grids-drive-51b-in-annual-utility-distribution-spending/528531/

4. Fast Facts about California Wind Energy (CalWEA Website)
https://www.calwea.org/fast-facts

5. Wind Speed Limits: Another reason for Electricity Shutdowns in California
https://www.heartland.org/news-opinion/news/wind-speed-limits-another-reason-for-electricity-shutdowns-in-california

6. The Photovoltaic Heat Island Effect: Larger Solar Power Plants Increase Local Temperatures
https://www.nature.com/articles/srep35070

7. MRO Electric – The most and least Power Outages in the United States
https://www.mroelectric.com/blog/most-least-power-outages/

8. Manatee County (Florida) will soon be home to the world's largest battery
https://www.heraldtribune.com/story/news/local/manatee/2021/08/12/fpl-nears-completion-worlds-largest-battery-solar-energy-project-manatee-county/8098241002/

9. Wind Speed Limits: Another reason for Electricity Shutdowns in California
https://www.heartland.org/news-opinion/news/wind-speed-limits-another-reason-for-electricity-shutdowns-in-california

10. California blazes a trail with 95 percent renewable energy
https://thehill.com/opinion/energy-environment/552218-california-blazes-a-trail-with-95-percent-renewable-energy

1: Identifying your needs – How much power do you need?

After you have identified your goals, you need to match your goals up with your actual needs. This means identifying what you need to operate – from refrigerators to cooling and hearting, communications, lighting, and communications or radios. You need to know how much power you need and when you need it. During the day you won't likely need as much lighting, but you may need fans or air conditioning. At night you will need lighting and possibly some heating. And you need to know what burns up your power. AC and heating are among the worst – each can quickly eat up 1500 watts/hour just for a small room.

What are you going to power?

This question will be unique to everyone. As an emergency radio operator my focus is on radio operation, running a small refrigerator, small lighting, and maybe small room AC during the day (Florida) or a small fan at night. Each family will have a different set of needs. For each system or tool that you will need, you need to understand exactly how much power that device eats. In my book "Small Power Planning" I go into more depth with explanations of how electrical power works and is measured. For our workshop I will be assuming you understand basic electricity and will give only a very brief explanation of what we are measuring

VOLTAGE: For most home appliances the power that they consume will be based on 120-volt AC power. This includes your television, refrigerator, computer, lighting, and your hair dryer. Think of voltage as the pressure at which electricity flows much the same as water pressure. If your needs include 240-volts, such as for a dryer or oven, then keep that in mind as many smaller power solutions will not provide 240-volts.

110, 115 or 120 Volts?
Which is right? We hear all three references often, what is the correct AC voltage level in the United States?

The answer is 120-Volts. References to 110-volts and 115-volts come from years past when our electrical standards were these. The proper household AC voltage is 120-volts or 240-volts for larger appliances.

CURRENT/AMPERAGE: The Current or amperage is the volume of electrons in electricity flowing through your wire. How fast current flows is based on your voltage, but amperage is the volume of the electrons. The larger and more energy hungry our devices are, the more current or AMPS they consume during a set time.

WATTS: The measure of WATTS is the combined measure of the voltage and amps flowing into a device. If a device is using 10 AMPS of current at a voltage of 120-Volts, then we multiply these two numbers together to get our WATTS being consumed each hour. So, a device using 10-AMPS at 120-VOLTS is using 10 x 120 Watts, or 1200-Watts. If that device is using 240-Volts of power (Dryer for instance) but still 10-Amps, then our wattage use is 10 x 240 or 2400-Watts. If on the other hand our 240-Volt device is only using 5-Amps, then our power consumed is 5 x 240 or 1200-Watts.

The WATTS measurement is an hourly measurement. Let's think of our microwave oven. If we have a 1500-Watt Microwave oven, then that Microwave oven will consume 1500-Watts running for one hour. However, I use my microwave oven no more than 10-Minutes a day. So how do I calculate how much power my microwave consumes in a day? Actually, it is a pretty simple calculation.

If I divide 1500-watts by 60 (minutes) then I get:

1500 (watts) / 60 (Minutes) = 25 (Watts per Minute)

So, I use 25-Watts/minute. Multiply that by the number of minutes I use it each day and I get:

25 (Watts/Minute) x 10 (Minutes) = 250-watts used over 25 Minutes

So, I use 250-watts per day for the microwave. That is what I figure into my daily wattage need for my Microwave. If we do this for each device that we are going to be using, we will quickly find how much power we *Think* we need. The table on the next two pages lists many common items we may wish to power. These numbers are only averages – you need to check the specific item you must know the amount of power it consumes. A 20 cubic foot high efficiency refrigerator for instance may average using 60 watts each hour. But you may have a smaller refrigerator that consumes more power.

Wattage Ratings of common electronics and appliances

In the chart on the next two pages, we are providing the *AVERAGE* wattage ratings of many around-the-home appliances you may find around the home and that you may need to power in an outage. Note that these are averages and specific power consumption will vary greatly among manufacturers and models.

Power planning is much like crystal ball reading in that you need to take think about what you need to power under different situations ahead of the event. During the summer in Florida for instance, you may want to consider how important AC is. For me, I can get through a hot day without air conditioning in the house, but if staying in my RV then AC is a necessity. For my older family members however, it could be life or death in the house. For this reason, I plan to have a 5,000 BTU Portable or window AC unit available to keep 1 small bedroom cool. That room could be a bedroom, or a sitting room we use for watching TV or playing video games.

What about during the winter? I'm in Florida – Our coldest days are not that cold – So I may decide to plan for nothing for heating. Layered clothes will work for me. But again – if there is someone sensitive in the home to temperature and you have extreme cold then a 1000-watt space heater in a bedroom or sitting room may just fit your needs. (Usually in the 30's in Central Florida is our extreme). But if you live somewhere that may be hit by extreme cold, you will want to consider other options. Some of these options may be electric, or others may be gas, or wood for a fireplace or wood burning stove. Your options will impact your electric needs and your plan.

Later in the book will provide a much more extensive list of over 300 items and their power usages. As stated again – you must look at your owner's manuals and energy guides for your specific equipment specifications. In chapter seven we will walk through performing an energy audit and walk you through an Excel worksheet that will help you with your own audit.

Table: Common power consumption

Item	Watts per Hour	Est Hours/Day	Watts/Day
Box Fan	200		
Ceiling Fan	120		
Central AC - 10,000 BTU	3250		
Central AC - 24,000 BTU	3800		
Tankless Water Heater	18000		
Water Heater – Electric	4500		
Well Pump - 1/3 1HP	750		
Window AC 10,000 BTU	1200		
Window AC 12,000 BTU	1500		
Window AC 5,000 BTU	600		
Blender	400		
Can Opener	170		
Coffee Machine	1750		
Dishwasher (1)	1800		
Chest Freezer - 15 cu. ft. (2)	45	24	1080
Sleep Apnea Machine (CPAP)	60	8	480
Refrigerator - 2.0 cu Dorm (AC) (2)	34	24	816
Refrigerator - 20 cu. ft. (AC) (2)	59	24	1416
Fridge -16 cu. ft. (AC) (2)	150	24	3600
Microwave (1)	1500		
Oven – Electric	2300		
Toaster	1400		
Toaster Oven	1500		
Clothes Dryer – Electric	5400		
Clothes Dryer – Gas	400		

Clothes Washer	512		
CFL Bulb - 100-Watt Equivalent	27		
CFL Bulb - 40-Watt Equivalent	10		
CFL Bulb - 60-Watt Equivalent	15		
CFL Bulb - 75-Watt Equivalent	20		
Compact Fluorescent 20 Watt	22		
Compact Fluorescent 25 Watt	28		
Halogen - 40 Watt	40		
Incandescent 100 Watt	100		
Incandescent 50 Watt	50		
LED Bulb - 100-Watt Equivalent	27		
LED Bulb - 40-Watt Equivalent	10		
LED Bulb - 60-Watt Equivalent	15		
LED Bulb - 75-watt equivalent	20		
Stereo Receiver	450		
TV - LCD (1)	150		
TV - Plasma (1)	200		
Video Game Console	150		
CB Radio - Transmit @ 4 watts	6		
Clock Radio (1)	10		
Curling Iron	150		
Electric Blanket (Twin)	63		
Hair Dryer	1500		
Ham Radio - Transmit @ 50 watts	75		
Radiotelephone – Receive	5		
Radiotelephone – Transmit	75		
Desktop Computer (Gaming)	500		
Desktop Computer (Standard)	200		
Laptop	100		
LCD Computer Monitor	100		
Printer (Non-Laser)	100		
Smart Phone – Recharge	6		
Tablet – Recharge	10		
Chain Saw - 12"	1100		
Circular Saw - 7-1/4"	900		
Circular Saw 8-1/4"	1400		
Drill - 1"	1000		
Drill - 1/2"	750		
Drill - 1/4"	250		

Resources and References:

- https://www.energy.gov/energysaver/save-electricity-and-fuel/appliances-and-electronics/estimating-appliance-and-home
- http://hes-documentation.lbl.gov/calculation-methodology/calculation-of-energy-consumption/major-appliances/miscellaneous-equipment-energy-consumption/default-energy-consumption-of-mels
- http://www.mea.coop/wp-content/uploads/2014/06/High-Bill-Packet.pdf

What if items I need to power are not in the list, or I want a more precise estimate?

Good question. Here are three solutions.

1. Look at the specifications in the owner's manual which should provide power specifications either in Watts used per hour, or maybe will list the voltage and current draw of the device. Then multiply Voltage x Amps to get Watts/Hour.

2. If you still have the ENERGYGUIDE label that came with your appliance, you can look at this guide for an estimated power use that can be broken down to a daily use figure.

3. Measure it yourself with a Wattmeter – A device that plugs into the wall and then you plug your device into it. The Wattmeter automatically tracks the power used over time.

Using the Energy Guide Label

The label to the right here is an example of the ENERGYGUIDE label that comes with most new appliances. This label gives us the estimated yearly operating cost as well as the estimated yearly electricity use amount. First you need to understand the yearly cost is an estimate and will vary depending on the cost of electricity in your area. But the kWh figure is a more accurate number and will be the value we are looking for here.

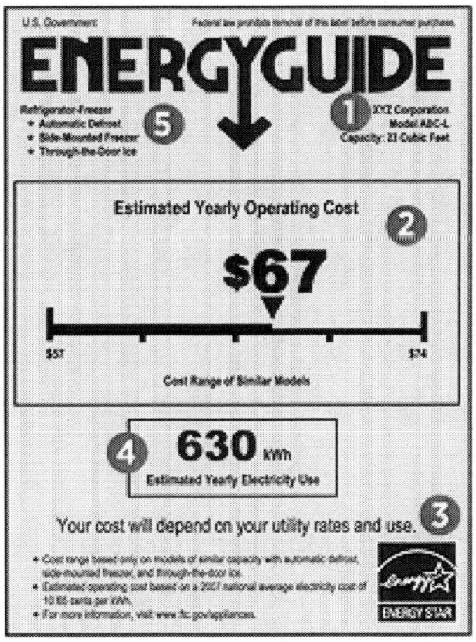

To calculate your daily wattage, use for this appliance, all we would need to do is the following:

kWh Yearly / 365 (Days in the year) / 24 (Hours per day)

630 kWh / 365 = 1.726 kWh each day (1,726 Watts)

Then for the hourly use divide by 24:

1726 / 24 = 71.91 Watts (72 Rounded off),

So, this appliance uses 72-watts each hour. Next, we need to calculate our daily use. If the appliance is a refrigerator which will always be on, we calculate the 72 by 24 to get a daily wattage estimate of 1728 watts (This is all for estimating). If the appliance is something else that we will use for 1 hour a day. Then our daily wattage estimate will be 72-watts.

The Wattmeter

The meter I use is the Poniie PN2000 digital watt meter available from Amazon for about $40. There are other manufacturers of watt meters, this just happens to be the one I have used and am familiar with. The meter as shown to the right comes with a short three-prong extension plus to help facilitate use of the meter. When you plug it in it is reset for tracking power use. Then over time the meter will track the number of minutes used and the amount of power flowing through it. Then all you need to do is divide the power use by the number of minutes, and then multiply that number by 60 (Minutes in an hour) to get the power use over an hour of use.

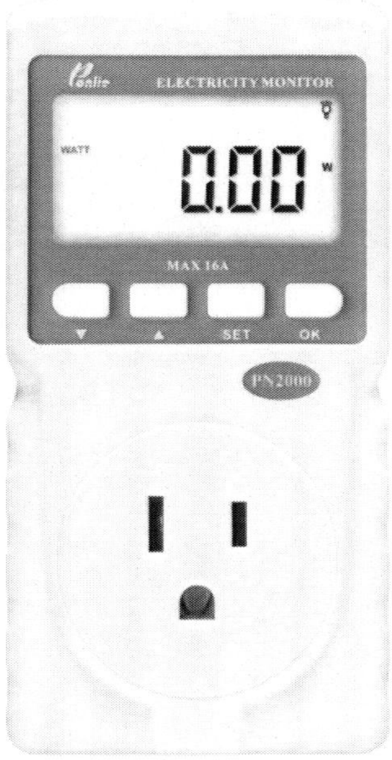

Poniie PN2000 Watt Meter

For many devices – lighting, televisions, computers, etc. you will always find a steady amount of power during use. (Computers and similar devices will change power consumption when entering sleep mode). For some devices such as refrigerators and freezers, and even microwave ovens, the power will rise, and fall based on the cycle of the device. For instance – When you first plug a warm refrigerator in, it will run hard for a few hours (or several hours) until it cools down to running temperature. Then, once at running temperature the refrigerator compressor will cycle off consuming very little power for a period. Usually, a refrigerator or freezer will only run about 1/3 of the time when cold. Because some devices can vary so much in how much power they consume, it is best to use a meter like this for an accurate measure.

My home refrigerator, about 20 cubic ft in size, consumes about 2.5-kw of power a day (2500-watts). I also have a small 3.2 cf refrigerator with a freezer that consumes 1100-watts of power daily. During a long-duration power outage I can power either easily with my gas generator, but when powering off my batteries at night then I would prefer only operating the smaller refrigerator if I am going to be running off-grid for a few days.

Note: *Harbor Freight also carries a "Kill A Watt" meter selling for under $30 that measures power going into an appliance also just like the Ponnie PN2000 meter discussed above.*

Cool and Warm rooms

For cooling a room during a hot day during a power blackout, you have two options. Either run the home AC unit which will drastically increase the size of generator you need to operate or consider being able to control the temperature in just one room. For myself and my kids, cooling is not a real concern, even in Florida. But I have elderly family members to consider also, so keeping at least one room cool can be life-or-death for them. After Hurricane Irma passed through Florida in 2017, over 40 deaths across the state were attributed to elderly residents in elderly care homes due to heat. Since then, the state has required nursing homes to install standby generators to keep their residents safe from heating problems.

So, what about cooling your home?

A good solution for a "Small Power" solution is to have a 5000 btu or so window unit that could be installed on-the-fly into a single room. If the room were a bedroom, convert the bedroom into a sitting room or have it as a safe room for anyone sensitive to the heat. This size window unit should consume as little as 500-watts for some models. This low wattage is easy for any mid-size generator assuming you are sizing for at least a 1000-watt to 3000-watt generator. Window AC units are more efficient with removing the heat from the room as the heat is dissipated outside of the room, out the window. For portable single-unit AC units that run inside the home and expel the warm heated air through a house and out the window – the problem is that the hose heats up that is running inside the room and dissipates heat back into the room reducing efficiency.

LG Portable AC Units can be easily installed and moved

Many homeowners will find a window unit to be a poor solution for many new homes using Central heat and air, but on a temporary basis and until normal power is restored this could be a good solution. If your home is 2-stories and your "Cool Room" will be on the upper floor, you may find a non-window cooling AC is easier to use. You still vent hot air out the window, but you only have a small vent to mount in the window instead of a bulkier window AC unit.

What about Heat in the Winter?

If you are going to be depending on Electricity for heating, then you can use a small room electric heater in the room which will consume from 1000-watts to 2000-watts of power. Alternatively, you could use something like a Buddy Heater which is marketed as a safe indoor propane heater running off the 1-lb propane bottles commonly used for portable camping stove or grill cooking. These will run for 5-8 hours safely. If you use any type of non-electric heater, always heed the warnings and instructions as some may require ventilation for use and may be deadly for use indoors.

If you do choose to heat with electricity keep in mind the cost during a grid outage will be high. Heating even a small electric heater will consume a lot of electrical power. If you are using a fuel generator this means the use of fuel that could be used for lighting, powering appliances, radios and communications, or other devices. If possible it would be best to avoid using emergency power for heating and look at alternative heating sources – Fire places, wood burning stoves, layered clothing,

Refrigeration needs and options

What about refrigeration and keeping food cool? Medications? Ice? Do you have any special items you need kept cold if the power goes out?

During the winter this can be as simple as keeping items outside. If it's cold enough you could make ice outside and use that ice in your refrigerator or freezer to keep your items cold. If your power outage is for a short period of time – four hours or less, the FDA says that your refrigerator and freezer should be fine without power. After four hours however many freezers and refrigerators will begin to warm up to the point of putting food storage at risk.

During a power outage, keeping your refrigerator or freezer cold can be easily accomplished from the use of a small generator. A large home-style refrigerator will consume somewhere between 75-watts to 200-watts per hour depending on the model and efficiency, while a small dorm-style refrigerator under 2.0 cf may consume roughly 30 to 35-watts per hour. In either case, a small fuel-based generator can keep either running during a power outage easily. If using a Solar generator, a mid-sized solar powered generator (500-watts to 1000-watts of battery) will keep a mini-refrigerator up and running for a few hours. We will talk more about solar generators and power stations later in this book including their limitations.

https://www.foodsafety.gov/food-safety-charts/food-safety-during-power-outage

Communications

What are your communications needs during a blackout?

Assuming you are planning for a blackout that lasts longer periods and not just a short duration rolling blackout of a few hours, you will at least need to have the ability to keep your phones charged. This is simple enough that it can be accomplished from your car, or even a small lightweight camping solar panel that has a USB output. This type of charging can be done easily using many solutions – Solar, from a vehicle, or fuel-based generator. But what if you need power for other types of communications?

Ham radio for instance, or an emergency CB or other type of radio? What about recharging a Handheld radio? All non-handheld radio types will run off a 12-volt battery directly, such as a smaller 20-amp/hour or similar battery (Handheld radios would require an adapter, but battery eliminators are available for most all small generators). The best small 12-volt battery would be a deep-cycle battery. Then plan for a method to keep the battery charged – Solar, or from a Fuel generator, or other method. If your radio runs directly off 12-volts, then keeping that battery charged is all you need to do. If your radio recharges from a USB port or through an AC tabletop cradle, then add a simple 12-volt-DC to 120-volt-AC power inverter and you are in business.

We will cover a solution using this method in a later chapter titled "Building your own Power Generator". This solar charger solution would work well as a minimal method to keep other AC powered devices. When we talk about building a power generator, we will cover sizing the generator and will discuss options such as use of the generator for larger devices.

Putting our numbers all together to assess our daily need

First – let's look at using a solar generator to get through the night with our appliances. The solar generator offers us access to power indoors at night with the safety of not producing fumes, operating silently (Except for a cooling fan maybe), and being secure as it is stored indoors with us. But – It is far from a perfect, or maybe even a good solution. I strongly believe having some form of battery based power solution is important for a wholistic approach – but it needs to be done properly or you will literally find yourself in the dark.

Let's say I have a 1000-watt Solar generator which can charge for each day at 500-watts/hour for 6 hours, and then during our off-peak sun hours we charge at a lesser amount – say 100-watts. During our daylight hours if we are consuming 100-watts each hour, we should be able to re-charge our batteries while still using our solar generator. At the end of our daylight hours let's assume our solar generator is 100% charged but needs to operate for the next 14 hours without any sun for charging.

Let's add a couple of small 10-watt LED lights in and the ability to recharge our phones at night. Our consumption will be estimated at 100-watts for 14 hours, so we will be consuming 1400-watts. Since we only have 1000-watts in our solar generator, then we do NOT have enough battery power to get through the night. We will lose power for at least 4 hours, probably a bit more. Our options are to supplement our power with something more such as another power source, expand our battery capacity, or simply go dark for the additional 4 hours.

Expanding our battery capacity is done either through the purchase of a larger battery, by adding an additional battery, or adding solar charging in such a way as our battery can be charged back up while being used. For some systems – Jackery Solar Generators for instance, they come in many different sizes. If you *Think* you only need a 500-watt system, perhaps purchase a 1-kW system would be better. For my Apex Solar Generator, I have two solutions to the limitation of a 1-kW of battery capacity. During the day I recharge the generator even when in use. I have over 400-watts of solar panels that keep it running and topped off most days. After the sun goes down, I have a 1.4 kW LiFeP04 battery that I can plug into the generator, via the Solar Charging ports – and charge it at a rate of about 140-watts/hour. My overnight use of the solar generator is about 100-watts each hour, so by morning it remains mostly charged.

Running your own Energy Audit using our Spreadsheet

In Chapter 7 we are going to work through an energy audit showing you how to notate each of the items that you want to have powered during a power outage. We have produced a multi-worksheet workbook and published it to our website at:

http://www.SmallPowerPlanning.com/EnergyAudit

This spreadsheet has a device list with over 300 variations of devices you may use throughout your home and throughout your workshop. After going through our example, you should be able to perform your own energy audit and then armed with that information you can make decisions on the size and types of backup power that works for you meeting both your power needs and budget.

2: Fuel Based Generators – Gasoline, Diesel, Propane and Natural Gas

Fuel based generators are the most cost-effective method of emergency power available when compared to Solar or Wind power options. A small generator for minimal power capabilities under about a thousand watts can be purchased for just a bit over a hundred dollars providing enough power for some lights and some charging capabilities. A good quality Inverter generator that produces AC power at about 2000 watts will set you back around $500. When compared to a solar based power solution, a gas or propane generator will set you back a fraction of the price of a comparably sized solar generator. Fuel generators also have several advantages over solar solutions when it comes to on demand power and being able to generate power during bad sun days or at night.

Larger Generators - Over 7-kW (Generally) – Standby Models

If you are looking for over 7-kW of power, you may be looking for a whole-home standby generator. These are available from several manufactures, such as Champion, Generac, Kohler, Briggs & Stratton, Cummins, and WinCo. These generators can be set to automatically turn on when the power fails providing you immediate power during a blackout. As these are "Large Power" solutions they are outside of the scope of our book – but here are some quick points for consideration:

- **Cost**
  These can quickly run into several thousand dollars for the larger generators after installation and with maintenance, however some whole-home generators that are smaller can be purchased and installed for less than $5,000 (Including labor, permitting and site preparation). Cost will depend on where you live – costs will include cost of an installer, items needed to connect to natural gas or propane, and permitting required by your local government office. These costs can vary widely based upon your location.

- **Portability**
  Whole home generators are NOT portable solutions. They are installed permanently at your home like how your AC unit may be installed at the side of your home.

- **Fuel**
  Your fuel options are Natural Gas, Propane (LP) or Diesel. With Propane or Diesel, you will need an appropriate storage tank for keeping the fuel at your home – an option that may not be available if you live in a deed restricted community. For Natural Gas – this requires a natural gas line to be installed running to your home. Natural Gas may be the only option in

many neighborhoods where you are prohibited from having a Diesel or Propane tank on site. Recall however that in the 2021 Texas freeze that natural gas stopped flowing in many areas leaving homeowners without power even with a home generator.

Small to Medium generators - Under 7-kW (Generally)

Small to medium generators cover a wide range of options from small (Noisy) gas generators under 1-kW in power capability up to the 6-kW to 7-kW size. The key factor in the smaller generator class we will cover is that these are portable generator solutions that can be pulled out and activated when needed but stored away when not needed. A portable generator gives you the option of grabbing the generator and taking it with you if you need to for operating at an alternate location. Portable generators these are often the preferred solutions for off grid living such as remote cabin homes or RV owners. A portable generator will meet our needs for a small power solution, but there are many factors in the generators which need to be reviewed to determine what type would work best for you.

Not all portable generators are created equal. Following are factors you need to consider when determining what type of portable generator will best fit your need.

Portable Generator Types – Open-Frame/Contractor

The less expensive portable generators are "Open Frame" or "Contractor" style generators. These are readily available and cost less than more advanced Inverter Generators, however they are MUCH louder and do not provide a clean alternating current power output. The noise may be a factor if you wish to keep a low profile. My 3500-watt inverter generator sitting 10 feet from my camper creates almost a quiet "Hum" barely heard from inside the camper (57db), while a smaller contractor-style open-frame generator can be heard loudly from several hundred feet away. If noise is a concern, stay away from these. If not, you can get more power from a less-expensive open-frame generator.

The other factor with most open-frame generators will be that most of these are "Modified" sine-wave output generators. This means that the AC sinewave is not smooth but rather jagged – a problem for some types of devices. If you are using one of these generators you need to confirm the equipment you will be using will operator properly on the output power. Some equipment will be damaged by a poor AC quality output from these types of generators.

The Open-Frame/Contractor style of generator does have its strengths. These are designed more for work use and therefore are often more rugged and designed for longer run times than other generator types. Some larger models also will have options for running diesel – not commonly available in smaller generators.

Portable Generator Types – Inverter and "Closed" frame generator

These are more modern styles of generators. Inverter generators are more expensive. They are popular with RV owners because of the quieter operation, and popular with many thanks to the smooth good-quality AC power they produce. The AC output power from these generators is safer for all equipment, including computerized devices and many types of sensitive medical equipment that could have problems with modified-sinewave power output from a non-inverter generator. The design of the inverter generator is such that the generator produces a non-clean AC signal which is converted into DC Power. The DC power is then used by an Inverter to produce a clean AC power source. The AC power is then referred to as a "Pure Sinewave" which has a "Total Harmonic Distortion" (THD) of 5% or less. The THD specification is important as it is the measure used to describe the quality of the AC power on many generators.

This chapter refers to Inverter generators and Closed frame generators. But it needs to be noted that one is not always the other. Some newer generators being manufactured are inverter generators with low THD specifications – but are open frame generators also. This means they have the quality AC output needed for sensitive electronics, but they do not have the lower noise levels as desirable for use by RV owners or anyone who wants to keep a low level of noise to avoid unwanted attraction.

One Generator or Two?

Many of the generators can be combined with another same-type generator for greater power output. For instance – Instead of purchasing a single 4000-watt generator, you could purchase two 2000-watt generators and combine them with a parallel kit. Advantages here include that the smaller generators each will weigh much less (About 50lbs vs. 100lbs) and that if you don't need the full power of a large generator, you could just run one. Another advantage is that if you need to perform maintenance, you could leave one generator running for power while changing the oil on the second generator or letting one rest while keeping the other going. (Yes – Generators need rest). If one of your generators fails, then you still have the second generator to depend on with 2,000-watts of power.

A Generator Parallel kit allows 2 generators to be paired together for a higher total output

Parallel kits are available for larger generators above 2000-kw. For my own Predator-3500 generator a third-party parallel kit will allow me to connect two of these generators together for 6000-watts run-time power or 7000-watts surge capacity. The newer Predator 7600w/9500w generator also has a Parallel kit to allow over 15,000-watts of power (19,000 watts surge).

Generator Run Time

How long do you intend to run your generator? Most smaller generators should be shut off after a run time of 8 to 16 hours (Depends on the Manufacturer and model). Some are good to run 24-hours or longer. If you are looking for a smaller gasoline generator you will have more run time restrictions than a diesel or many propane generators. For this reason – two generators connected through a parallel kit would allow you to run one or the other generator while the other is taken offline for rest or maintenance. ALWAYS follow the manufacturer's recommendation for runtimes to keep your generator operating properly.

Gasoline generators should be shut off after several hours of use for rest. As a rule of thumb, it is best not to keep gasoline generators running more than 12 to 18 hours (Depending on the specific generator) while a propane generator, which is cleaner burning, could operate for 150-200 hours if continuous fuel were available (Again – depending on the generator). Always when running generators for long periods of time you need to monitor the oil levels in the generator and refill as necessary.

Larger generators such as standby generators which are permanently installed for backup power can be operated for longer periods of time – some for about 48 hours, while others may be able to operate for 500 hours (About 21 days). If a long runtime is critical, in this case you should look at a larger standby generator running off Diesel, Propane or Natural Gas.

Fuel options

What type of fuel works best for you, and what type would be problematic? Whether you are looking for a larger standby generator or are planning to use a smaller generator, the type of fuel you are going to be using is an important consideration in your selection. Following is a breakdown of the fuel options with pros and cons of each:

Gasoline

Gas is the most available fuel and provides more power output than propane when used in a smaller portable generator. When using gasoline, use of ethanol-free gasoline will create fewer problems for your generator and keeps it in better condition. (Ethanol absorbs water when stored which is a problem). Benefits to Gasoline include:

- More available than other fuels for portable use
- Though less energy by volume than Diesel, it has more energy than Propane or Natural Gas.
- Can be easily stored in relatively inexpensive storage "Cans" – though it will break down sooner than Propane or diesel in storage. (Natural Gas is not stored at home).

One drawback from gasoline is what makes it the most available – Larger demand. If you are in an emergency, it may be more difficult to obtain gasoline. For Hurricane planning in my area this is solved by getting gasoline early before there is a run at the pumps and keeping at least 2-3 extra 5-gallon cans available. Another drawback of Gasoline is shorter storage life. Gasoline requires an additive for longer term storage over 6 months.

Propane

The next most popular fuel for small generators is Propane. Propane has an advantage over gasoline in its ability to be stored without degradation over a longer period and it is a cleaner burning fuel over gasoline and diesel. Here are the major advantages of Propane:

- Stores longer – Much greater longevity
- Does not have to be "Pumped" – During a power outage propane tanks can be filled without electrical pumps.
- Cleaner burning fuel – As such, this will reduce maintenance on your generator
- Spill-Free – Propane is a gas at outdoor temperatures and will dissipate as a gas, not a liquid.
- Not affected by cold weather like Gasoline or Diesel. A better fuel option in cold conditions.
- Propane is stored locally – so if you use larger tanks for storage you will not have to worry about problems with distribution lines such as Natural Gas.
- Though less energy than gasoline or diesel, Propane has more than twice the energy per cubic foot when compared to Natural Gas.
- Generally, propane is less expensive than Gasoline or Diesel, but with less energy density this will be offset and in the long run may very well be more expensive for fuel based on energy produced.

Propane Drawbacks include:

- Less energy dense than gasoline – Propane will burn faster and produce less power than gasoline or diesel.
- Less available than gasoline or diesel
- More expensive based on energy produced

Note: In many areas – if your HOA does NOT allow for installation of a large propane tank, BUT they have allowed or not restricted large tanks in the past, then this allowance in the past may well nullify any restriction they could enforce. This depends on where you are and local laws.

Diesel Fuel

Diesel fuel is an excellent fuel for generators, though not as commonly used by smaller generators. As a store of energy, Diesel stores more energy per unit than gasoline, propane, or natural gas. When shopping for a generator you will find that Diesel is used only for medium or larger generators. Following are some of the advantages for using Diesel:

- Diesel fuel stores more energy than other common fuels – between 10% and 15% more than gasoline.
- Less flammable and better for storage than gasoline
- During an emergency even such as a hurricane or natural disaster, a run on gasoline may result in gasoline shortages. At these times diesel fuel may be easier to obtain.
- With no carburetor or spark plugs a diesel generator will require less maintenance.
- Diesel-based generators will often run longer periods without rest – Gasoline generators will require rest periodically.
- Diesel, along with Propane, will be preferred choices of fuel for larger generators.
- Though in the past considered a "Dirtier" fuel than gasoline, newer generators will burn much cleaner than older generators.
- Newer Diesel generators are also quieter than they have been in the past, though not as quiet as an inverter generator.
- Diesel generators are more rugged and longer lasting generators.

Drawbacks to Diesel generators include:

- Diesel generators are larger due to the design of how the engines work. Larger equates to heavier, and bulkier – harder to move.
- More expensive than gasoline generators. A 5-kW diesel generator for instance starts in the $3k to $4k price range, where a Gasoline based generator capable of similar output can be purchased for under $1k

Natural Gas

Natural Gas is a very popular fuel choice for larger stationary and home-based generator systems. This fuel is delivered to a home or business through installed natural gas lines through your local utility. Due to the type of generators that commonly use natural gas being larger stationary generators, we will not be discussing it much in this book. Natural gas is an excellent fuel except in situations where delivery could be interrupted. During the February 2021 deep freeze in Texas, much of the power failure was due to the freezing of natural gas lines. For a stationary generator for whole-home backup strategies and in a community not allowing for propane storage in large tanks, natural gas will be the preferred fuel choice.

Natural Gas is not an option for portable generators, but rather for standby generators. As it is not a fuel option for the generators that we are focusing on we will not discuss it further.

Important Note about Generator Comparisons:
This is raw power output comparison – But not necessarily a similar specification comparison. For instance – Generac's XD5000E Diesel generator is more expensive – 3x more than a Gasoline generator capable of similar output. But – The Generac generator is a near pure-sinewave generator with a THD rating under 6% and uses Diesel fuel with a fuel tank size of 12 gallons allowing up to 32 hours of run time based on 50% load. So, while power output may be similar in watts to a gasoline generator at a much lower price, this generator is not an Apples- to-Apples comparison as it has features and capabilities beyond than similar wattage rated gasoline generators.

Security Concerns – Generator Noise

When it comes to noise – an enclosed Inverter style generator will win hands down for quieter operation. Noise for generators is measured in decibels (dB). Larger contractor-style and open-frame generators will operate in a range of 70 dB to 100 dB (Loud – About as loud as a gasoline lawn mower up to heavy city traffic). Enclosed Inverter generators will range from 50 dB to 70 dB (Normal speech to Vacuum Cleaner level) while open-frame inverter generators will operate in the 70db range and higher.

When it comes to security – the quieter the better. If you are running a generator at night during a power blackout, the noise will be an attraction you may not want. If you are operating an Inverter-style generator, this alone will be much quieter than a contractor-style generator – quieter enough that running it from the back of your house may block it from even being heard.

For my own use, I run an Inverter generator for camping and will use it during blackouts for short-term power. But at night – if I feel that I need to maintain a greater level of silence, I will switch over to a solar generator capable of powering limited lighting, television, and refrigeration. The solar generator works in complete silence with no fumes and is safe to operate indoors. The next day I can again operate the gasoline generator for appliances and even use it to recharge my solar generator or use solar panels to recharge the batteries in the solar generator for use the next night if necessary.

For smaller inverter generators, ranging from 1600-watts up to about 4000-watts, there are many options from several manufacturers. Honda, Yamaha, Wen, Predator, Westinghouse, Champion, and many others produce fine quality and capable generators.

A Short List of Inverter-based generators for review

Let's look at the specifications of many popular inverter generators. In this list are a partial sampling of many popular generator models, but for most of these manufacturers only a few models are listed from those manufacturers. Specifications to look at include not only wattage size, but also the type of starting system, fuel, decibel level, weight and if the generator supports 240-volt output. Not all generators excel at every category or will be the "Perfect" solution but depending on your needs many will fit your needs better.

Generator	Run Watts	Max Watts	Start Type	Fuel	dB	Frame	Low-Oil	240-V	Weight	Tank Size	Price
Champion 100520	7000	8750	Electric/Pull	Gas	72	Open	Yes	Yes	149	4.2 Gal	$1,199.00
Champion 100574	3500	4000	Pull	Gas/LP	64	Open	Yes	No	82	2.9 Gal	$629.00
Champion 100899	1850	2500	Pull	Gas/LP	53	Closed	Yes	No	47.6	1.1 Gal	$614.00
Champion 200992	3650	4650	Pull	Gas	61	Closed	Yes	No	101	2.3 Gal	$799.00
Champion 100263	3100	3400	Electric/Pull	Gas/LP	59	Closed	Yes	No	98	1.6 Gal	$1,049.00
Champion 100519	5000	6250	Pull	Gas	75	Open	Yes	Yes	121	4.0 Gal	$949.00
Honda EU1000i	900	1000	Pull	Gas	59	Closed	Yes	No	28.7	.55 Gal	$899.00
Honda EG2800i	2500	2800	Pull	Gas	69	Open	Yes	No	66.6	2.1 Gal	$1,149.00
Honda EU3000iS	2800	3000	Electric/Pull	Gas	58	Closed	Yes	No	134	3.4 Gal	$2,299.00
Honda EU2200i	1800	2200	Pull	Gas	62	Closed	Yes	No	46.3	.95 Gal	Unknown
HF Predator 2000	1600	2000	Pull	Gas	65	Closed	Yes	No	47	1 Gal	$549.00
HF Predator 3500	3000	3500	Electric/Pull	Gas	57	Closed	Yes	No	99.2	2.6 Gal	$849.00
HF Predator 8750	7000	8750	Electric/Pull	Gas	75	Open	Yes	Yes	141	5.9 Gal	$1199.00
HF Predator 9500	7600	9500	Electric/Pull	Gas	67	Closed	Yes	Yes	257	6.8 Gal	1799.00
Generac XD5000E	5000	5500	Electric/Pull	Diesel	(A)	Open	Yes	Yes	247	12 Gal	$3,999.00
Generac GP3500iO	3000	3500	Pull	Gas	(A)	Open	Yes	No	75	2.37 Gal	$789.00
Generac IQ3500	3000	3500	Electric/Pull	Gas	(A)	Closed	Yes	No	109.1	2.6 Gal	$1,049.00
Generac GP2500i	2200	2500	Pull	Gas	(A)	Closed	Yes	No	48	1.0 Gal	$599.00
Yamaha EF1000iS	900	1000	Pull	Gas	57	Closed	Yes	No	27.9	.66 Gal	$779.00
Yamaha EF2000ISV2	1600	2000	Pull	Gas	61	Closed	Yes	No	44.1	1.1 Gal	$849.00
Yamaha EF3000ISEB	2800	3000	Electric/Pull	Gas	60	Closed	Yes	No	154.3	3.4 Gal	$2,215.00
Yamaha EF6300ISDE	5500	6300	Electric/Pull	Gas	64	Closed	Yes	Yes	200	4.5 Gal	$3,650.00

Table 2-1: Various popular Inverter Generators

(A) Generac does not provide noise level estimates for their generators

Inverter or Contracting generators – Pricing Comparisons

First – I'm not trying to sell you on needing to get an Inverter generator, or a generator with a pure-sinewave output. But – I do want to stress the need to make sure any sensitive devices you will be powering will not be damaged by the power quality from a less expensive generator that does not provide clean power.

Below we are looking at the line of Harbor Freight generators as of Spring 2021. As you will see from the list of generators, the price difference between the modified sinewave style generator and the inverter style generator is pretty great. If your needs do not require the clean AC sinewave of an Inverter generator or a generator with a low level of Harmonic Distortion (Measured as 'THD'), then you can easily get more power for the same cost by choosing a modified sinewave model.

	Generator	Sine wave	Run Watts	Max Watts	Starting Type	Fuel	dB	Frame	Low Oil	240-V	Weight	Tank Size	Price
1	HF Predator 2000	Pure	1600	2000	Pull	Gas	65	Close	Yes	No	47	1 Gal	$549.00
2	HF Predator 3500	Pure	3000	3500	Electric/Pull	Gas	57	Closed	Yes	No	99.2	2.6 Gal	$849.00
3	Predator 4000	Modified	3200	4000	Pull	Gas	72	Open	Yes	Yes	99	4 Gal	$349.00
4	Predator 4375	Modified	3500	4375	Pull	Gas	72	Open	Yes	Yes	103.5	4 Gal	$429.00
5	Predator 6500	Modified	5500	6500	Pull	Gas	74	Open	Yes	Yes	165	8 Gals	$549.00
6	Predator 8750	Modified	7000	8750	Electric/Pull	Gas	76	Open	Yes	Yes	183	6 Gal	$599.00
7	Predator 9000	Modified	7250	9000	Electric/Pull	Gas	76	Open	Yes	Yes	200.8	8 Gals	$749.00

Table 2-2: Various popular Inverter Generators

In the prior list from Harbor Freight, let's compare two models that are somewhat close. We will compare the Predator 3500 Inverter generator to the Predator 4000 Contractor generator here.

Generator/Model	Category	Running Watts	Sinewave	Fuel	Weight	dB	Price
Predator 3500-Inverter 	Inverter Generator 11 Hour Runtime @ 25% Load (.23 Gals per Hour)	3000 (3500 watts Surge)	Pure AC 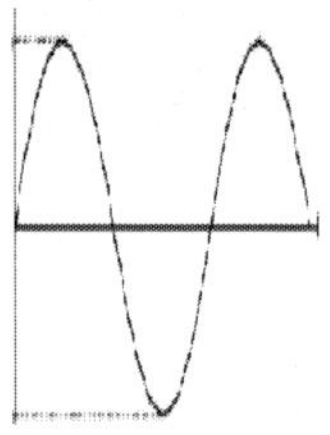	Gas 2.6 Gallons	99 lbs	57 dB	849
Predator 4000 	Contractor Generator 10 Hour Runtime @ 50% Load (.4 Gals per Hour)	3200 (4000 watts Surge)	Modified 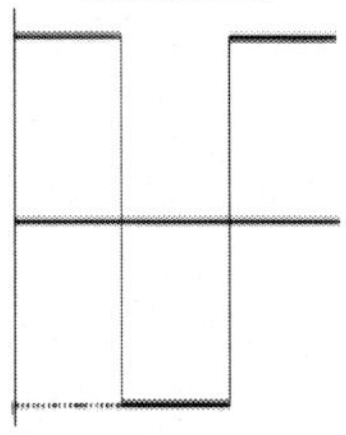	Gas 4.0 Gallons	99 lbs	72 dB	349

Table 2-3: Comparing two similar generators, Inverter vs. Non-Inverter

In Table 2-3 on the prior page, we see the specifications that most stand out in comparing an Inverter to a Non-Inverter generator.

Looking at the generator category – we can see that the minimal burn on fuel for an Inverter generator is going to be less than that of a non-inverter. This is because the Inverter mechanism in the generator can throttle or idle down the motor allowing a slower burn of the fuel. In this example, even though the inverter generator has a smaller fuel tank, we see that the ability to throttle down will allow us to burn fuel slower and lead to a more conservative generator. There will be a lot of times when a 3000-watt generator does not need to provide that much power. At full load both generators will burn fuel at a similar rate. So, for fuel efficiency, the inverter generator wins.

Next, we look at the sinewave out. The inverter generator provides a clean AC signal out – cleaner than the modified sinewave from the contractor generator. This is a win for any sensitive electronics needing this clean power, and any equipment that has an inductive load, such as items with inductive motors. Devices such as radio gear running from a modified sinewave power source may also pick up noise from the power lines or even interference heard on the radio. Again, the Inverter generator wins here.

On Fuel – Both of these generators run off gasoline only. The Contractor does have a larger fuel tank and if the load will be steady and high where lower throttling does not happen, then the additional fuel and longer runtime of the contractor generator will be a win for that type of generator. *(Even Gas-Only generators can run off Propane – This requires a third-party conversion kit such as from **Century Fuel Products**.)*

https://centuryfuelproducts.com/generators/conversion-kits

The next characteristic with a significant difference is the dB noise rating between the generators. The Inverter generator is enclosed, plus with the ability to throttle down the motor, makes this generator much quieter at about 57 db. This also make the inverter generator the preferred generator when either RV'ing (Quieter around others) and if you want to keep others from being aware you are running a generator – the noise can create unwanted attention.

The last category is cost – and this is a big category. In this case, the inverter generator is over twice the cost of the non-inverter. The Inverter generator running $849 is over twice the cost of the Contractor version of the generator running just $349. In fact, for the same $849 cost of the inverter generator you can purchase the contractor style generator from HF putting out 7250 running-watts (9000-watts peak) for $749.

As mentioned earlier, I am only using the Harbor Freight Predator generators as a comparative example. I am not making a recommendation of these generators over any other brand or models. The top-of-the-line manufacturer of generators is the Honda brand. The comparative generator from the Predator 3500 coming from Honda would be the Honda EU3000iS Inverter generator with similar specifications. With the price point of the Honda being over $2,500 however at the time of this writing, you can understand the attraction to the lesser expensive generators. Honda however has the best reputation for quality and will also enjoy a wider base of service providers.

Last Minute Generator Update – Harbor Freight 2021 New Releases

As I am preparing to publish this book, Harbor Freight Tools (HFT) has announced the release of five new generators that would be of interest for backup power in multiple categories. Capabilities as well as pricing are of interest enough to include for a last-minute mention for this book. As mentioned earlier, I am not specifically advertising Harbor Freight or receiving sponsorship from any manufacturer I am mentioning.

- **Predator 9500 Enclosed Inverter Generator - $1799**
  This is the new tp-of-the line inverter generator producing 7600-watts running, 9500-watts surge power. Running at 67db this generator is louder than HFT's smaller inverter generators, but quieter than similar open-frame generators. As an inverter generator the fuel burn rate is also lower compared to a non-inverter generator. This larger generator provides also 240-volts output as well as traditional 120-volt output and weighs in at a hefty 257 lbs making it less on the portable side but does include a wheel kit to facilitate moving.

With a 6.8-gallon fuel tank HFT is touting over an 18-hour run time for the generator making this a great large-size power option that can provide power for your home or large needs during a blackout.

- **Predator 1800-watt Gas powered Open-Frame non-inverter generator - $279**
  HFT's new entry level generator is an 1800-watt open-frame and non-inverter generator providing 1400-watts of running power with a 1.6-Gallon fuel tank touting over 9 hours of runtime at 50%. This generator is replacing the small 700/900-watt tailgater.

- **Predator 1100-watt/1400-watt Inverter Generator - $429**
  This is HFT's new entry level Inverter generator providing 1100-watts of running power, 1400-watts of surge power. Weighing in at just 33lbs, this is also the lightest of the generators now produced by HFT.

If noise level is not a concern for your needs, Harbor Freight Tools is also introducing two open-frame Inverter generators which will offer benefits of clean AC power with a lower price point. These include:

- Predator 4400-watt Open-Frame inverter generator - $599
  This generator is a 4400-watt surge, 3500-watt running rated generator with a noise level of 67db. (Similar to the new 9500-watt Inverter model, but louder than the 57-db rated Predator 3500 and Predator 2000 generators. The 3.2 Gallon fuel tank offers a run time of 16-hours at a 25% load.

- Predator 8750-watt Open-Frame Inverter generator -$1199
  This open-frame generator provides 7000-watts of running power, 8750-watts surge power at 75db. The 5.9 Gallon fuel tank offers a run time of 15-hours at 25% load making this a louder but great portable solution for a larger need such as a home or large-power requirement.

3: Small Solar Power – The Solar Generator

If you are looking for a power solution that will not depend upon fossil fuel, or that runs silently, or a solution that is fumeless allowing it to be run indoors, then a Solar generator is the solution you are looking for. A solar generator is a not actually a "Solar" generator. Rather it is a portable battery powered energy source that provides electrical power in the form of DC power that can be used to power 12-volt or USB devices, and AC Home power that can power home AC devices. Where the term "Solar" comes from is the usual design that these devices can recharge from solar panels. We are going to talk about Solar generators in detail, but first I want to discuss whole home solar solutions for contrast.

Whole Home costs and realities (Large Power Solutions)

Whole home solar extends beyond our discussion in this book because this is what we would consider to be a "Large Power" solution. For purposes of being thorough we will touch on it here, but our goal is not going to cover it in depth. In the chapter for Rolling blackouts coming up we will review several battery options that can provide large scale power in your home.

Grid Tied Power

There are essentially two categories of whole home solar that are available to homeowners for installation. The first is a grid-tied solar only solution. With this solution, you will have solar panels installed at your home location for the generation of electricity during sunlight hours. This power that is generated can be used to either reduce or even eliminate your power bill. Electricity that you generate during the day can be "Stored" in the grid. Then at night your power comes from the grid, using much of the power you were credited for during the day earlier. Basically, the electric company is crediting you for power you generate during the day and allowing you to use that power at night.

Remember that credit you received from the electric company for excess energy generated during the day that you did not use? Well at night or at times when you cannot generate enough electricity (Rainy days) you eat into that credit. At the end of the month if you have generated enough electricity you should have either a greatly reduced electric bill or no electric bill. This type of solar is great – you are reducing or eliminating your monthly electric bill. But – You are not storing electricity. All electricity that is excess is going back into the grid for use by a neighbor. The problem with this however is that during a blackout you are left with all these great solar panels, but because you are tied in with the grid and the grid is down, so are you. During the 2017 hurricane season in Florida, Hurricane Irma left over 6 million Florida homeowners without power for over a week – many of which had solar panels on their roofs with no way of using that electricity.

Solar Panels with Batteries

The next category of whole home solar includes the same panels for the home as we use in a grid-tied system but adds to the solution a battery infrastructure and battery charging that is capable of not only storing the electricity being generated, but also is able to feed that power back to the house. This system uses powerful electrical inverters which convert that stored DC electricity into household AC power that is used by all your appliances. The battery storage capacity may be less than what you normally use during the day, so you may have to budget your electrical use during a power outage, but you do have power. This means you can run your lights and TV, but you may have to budget your water heater or go without Air Conditioning. But – Not to worry – if you are willing to put enough money into your system, you can power everything. During a non-blackout, the system will feed your house partially, pulling additional power from the grid. But – the system will still greatly reduce your daily energy costs.

Whole Home Solar Costs

So how much does this all cost? This depends on the size of system you want installed. Most home solar systems will run into the 20k price range, and as you add batteries and inverters will shoot up from there. Most likely you will be in the 20k to 40k range for installing solar power to the house. There are great cost savings to this however which will reduce or eliminate your electric bill so I am in no way going to discourage this – in fact I believe that due to grid reliability dropping and risk of disasters seeming to be growing, I think it is a good idea for homeowners to take steps to insure their able to remain powered and generating power during what could be long blackout periods.

Whole home costs are a significant drawback to this type of large power system, but also is the permanence of the installation. What happens if you live in an apartment? Or you need to evacuate to stay with someone else? If you need to head to a shelter, or if the situation requires you evacuate your home for days or weeks? Whole-home systems will be of no benefit in these situations. In our next chapter we are going to look at rolling blackout solutions which cover some of these systems.

Small Solar Power

If you come into a situation that requires you to leave the area where you are, sheltering or staying with someone else – a portable solution will allow you to take some of your power generation capabilities with you. This is where the solar generator comes into play and where we will focus on discussion for the rest of this chapter. As we mentioned in the opening of this chapter – a "Solar Generator" is a portable solar power system in a single box – except for the solar panels. It is miniaturized and capable of providing a variety of power types in a package that can be kept with you. Small generators weigh as little as 6-lbs and can power small AC devices or used to charge a phone. These can be great for camping, traveling, or hanging out in an evacuation shelter and power devices you need while you wait to go home.

This 6-lb lightweight generator will not store much power, but you can greatly increase your power capabilities with larger models of generators without increasing the weight or size too much (Depending on the manufacturer). I have a 25-lb solar that I can carry that can put out up to 1500-watts of power and power my 30' travel trailer, minus the air conditioning. I can also power ham radios and emergency radios, emergency lighting, my refrigerator, and even connect to 500-watts of solar panels for recharging the generator during the day – even while I am using it. It also supports external batteries or input power from an external DC power source.

The Solar Generator can be a device that you purchase, or one that you can even build. We could walk into Harbor Freight (One of the few retailers that sells solar panels off the shelf) and walk out with about $400 worth of parts to build one right now. (In a later chapter we will lay out two build plans to put together small solar generators in the 200-500-watt range using a plastic Ammo box or a larger Home Depot toolbox, or just laid out on the floor. Let's look at what actually makes up a solar generator so you have a better understanding of the magic in the box, then we will review some off -the-shelf models you can purchase along with specifications to look for.

Components of a Solar Power System

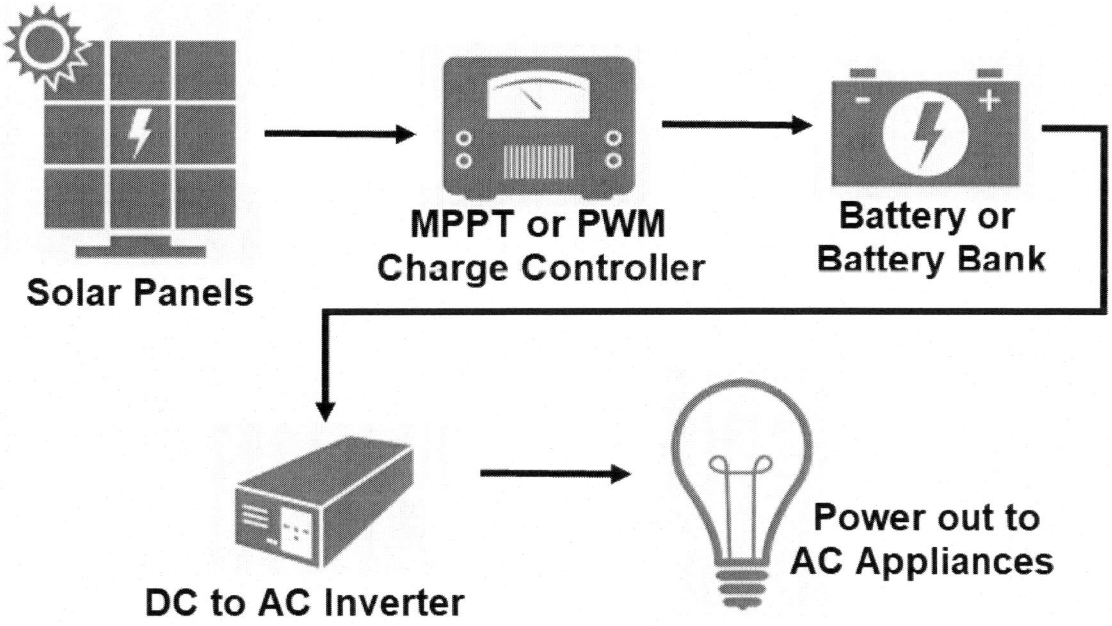

In the diagram above we have the general components of a Solar Power system. This applies to both a full size whole-home system, or a small solar generator. There are components of course that are minor components that are not included, such as shunts, wiring, bus bars and other minor items.

Solar Panels: The solar panels are always external in a solar generator. These can be one or multiple panels, from small portable suitcase or folding panels for portability, or larger 100-watt rigid or flexible panels, or even larger sized panels designed for home or commercial mounting. My RV panels that are used to charge my solar generator include a large 340-watt panel in series with up to three 100-watt panels. Simultaneously I have four 100-watt panels that I can carry and lay out as needed to charge my solar generator while mobile or tent camping.

Charge Controller – MPPT or PWM: The charge controller takes in power from the solar panels and regulates that power to charge your batteries. In my RV system, I pull in about 60-volts from my panels in a series/parallel configuration, and the MPPT charge controller I use regulates that down to 13.8 volts for my lead-acid battery bank. The charge controller I use is capable of charging Lithium or Lead-acid based batteries though many charge controllers can only charge specific battery chemistries. MPPT Controllers are superior and more efficient than PWM charge controllers. In a solar generator the charge controller is integrated into the generator. For larger systems, such as for an RV or home, the charge controller will usually be a stand-alone device mounted on a wooden board or panel.

Battery or Battery Bank: The Batteries are the energy storage component of the solar generator. In a Solar generator the battery is integrated into the solar generator and is most often a Lithium based battery. In an RV or home solar system the battery is a single battery, or a bank of multiple batteries made up of Lithium or Lead-acid batteries. Lithium is usually considered the superior battery type able to store much more energy, weighing much less, and able to be discharged to nearly empty and still be fully charged. Lithium batteries also are superior in the number of cycles that they can be recharged lasting many times longer than lead-acid batteries. (Typical charge cycles for Lead based batteries is around 500, while up to and over 2000 for Lithium.)

DC to AC Power Inverter: The power inverter is used to convert electrical power from the DC batteries to Alternating AC power which is compatible with home appliances such as your television, lighting, refrigerator, and other household items. The inverters come in various sizes and are categorized as "Pure Sinewave" or "Modified Sinewave". The modified sinewave puts out a dirtier AC sinewave with a THD (Total Harmonic Distortion) greater than 5% which can be harmful to sensitive electronic devices containing a microprocessor, but these are also less expensive. The Pure Sinewave models put out a much cleaner sinewave with 5% or less THD. For a small inverter the cost difference is not too much – but as we increase sizes costs do jump. (A 300-watt Pure Sinewave inverter will run about $50 – about the same as a 750-watt Modified Sinewave inverter). The power inverter will have one or more electrical outlets on it. For our solar generator, the inverter will be wired to one or more outlets on the case of the inverter which can be 120-v AC, 30-Amp RV Plugs, 12-volt or even USB power outlets. Below is an image of the front panel of the Inergy Apex Solar Generator showing multiple types of output ports it has available on it. On the Apex we have a 30-Amp output, (6) 120-volt outputs, (2) 12-v DC Cigarette outlets, (4) USB Output ports, and (2) 12-v DC Basecamp light power ports.

Inergy Tech's Apex Solar Generator – Has multiple outlets for AC and DC Power out, up to 12.5 Amps. Most similar products include 120-v ports, multiple USB ports (A & C), 12-V Accessory plugs, and some include other plugs such as RV-style plugs and accessory plugs for lights or other devices.

Power out to AC Appliances: Here we are referring to Alternating Current appliances in the home. With most solar generators as with the Apex in the image above, we have output going to AC 120-volt outlets, 12-volt, and USB devices. Solar generators are designed as simple, compact and portable units so usually you will just plug directly into them, but some are large enough that they can directly interface into your home or RV to allow a single connection to power some power outlets. In my RV for instance, I plug the shore power directly into the RV plug of the Inergy Apex and then power is fed into the entire trailer. I cannot power large items such as my Air Conditioner, but all my trailer outlets are energized and if I stay under the maximum power output of the generator, I can use those devices. This includes both my Refrigerator and my Microwave oven. If though I turn on too many devices which overload my solar generator, then the generator shuts down.

Solar Generators Off the Shelf

Solar generators come in many sizes and models with many features – There is certainly a solution that would fit your needs. But from who? And what specifications? We are going to review first many of the specifications of a Solar generator, and what works best, as well as what to stay away from. Then we will look at several manufacturers that have shot to the top of many recommendation lists.

Spec: Battery Chemistry

The battery chemistries used in the solar generators are either based on a Lead-Acid battery type or a Lithium battery type. There are several types of each, but each family has similar characteristics. Lead-Acid batteries have been around for ages, but they don't hold up in comparison to the newer Lithium technology and Lithium is coming down in price. With Lead acid batteries – the battery rating is NOT the amount of energy available for use. To keep from damaging the battery you should use no more than 50% of the battery charge. More and you will shorten the battery life. The battery's voltage for a lead-acid battery drops to about 12.05 volts when it is at 50% charge. Below that and you will damage the battery.

With a Lithium battery you can use between 90% and 100% of the energy the battery is rated for thanks to the "Battery Management System" (BMS) built into most systems. So, a 1200-watt hour Lead-Acid battery only has 600-watt hours of energy available. The same size lithium has between 1080-watt hours and 1200-watt hours available (Varies by manufacturer). Plus, lithium weighs in at about 1/3 to 1/2 the weight – important if you are either carrying, or if you are building a battery bank somewhere such as in an RV.

How long the batteries will last is an important characteristic as well. Lead-Acid batteries last, in general, between 500-1000 charge cycles before they degrade in quality. With lithium batteries, you will see between 2,000 and 5,000 charge cycles before degradation, and even then, they still hold a very high percentage of their energy storing ability. Looked at with regards to battery storage and overall life – the lithium batteries come out much cheaper than their lead-acid counterparts.

Spec: Charge Controller – MPPT or PWM
Most new solar generators will come with an MPPT charge controller – also known as a *Maximum Power Point Tracking* charge controller. These are more programmable, support more configurations, usually support lead-acid *and* lithium batteries, and are more efficient than their PWM counterparts. They are also more expensive to purchase or to integrate into the solar charger bringing the cost up. The higher efficiency is important to ensure faster recharging of the solar generator or batteries.

The PWM charge controller (Pulse Width Modulation) is an older technology for charging batteries – but still works well. These are less expensive charge controllers which are more specific for their use – often (but not always) designed for a specific battery chemistry. PWM charge controllers are known for being less efficient and being able to support only smaller configuration (85% to 95% efficient). If you are looking for a smaller or less expensive solar generator, a PWM controller will still work for you, it will just slow your charging due to efficiency loses. In both designs we talk about later for a home-built solution we use PWM controllers.

Spec: Solar Input, Charging Rate and Overall Charging Rate
The charging rate is how fast the solar batteries can be recharged. For a solar generator this is an important specification when it comes to panels – you will want to be able to recharge your generator as quickly as possible or with the maximum number of panels possible. Remember – at night there is no charging from the sun, so you want to collect as much during the day and be as close to fully charged as possible at sunset when you need to start budgeting your power. When I purchased my Apex the maximum charge rate was up to 500-watts in through external panels. The next most popular was at the time the Goal Zero which could only charge in the 200-watt range. (Since both have increased with Goal Zero introducing several much more powerful models).

In addition to charging from solar – you can also charge from the AC wall outlet. If you are in an area with rolling blackouts it may be that you don't even need solar panels. Charging from the wall may be enough to carry you 4 or 5 hours during the blackout.

Spec: Power Inverter, Size and Quality of AC Power
Another critical factor comes from the Power Inverter. First – Is it a Modified sinewave or a Pure sinewave. This is <u>CRITICAL</u> for some types of equipment you will be powering. Some devices that have microprocessors in them will be sensitive to dirty AC power. Devices that have inductive motors or inductive loads will have difficulty with poor quality power. Some devices will burn up and fail – and yes, this has happened to me. A pure sinewave inverter will add to the price of the device, but it will ensure you can run anything on it.

The other factor with the power inverter is the size – how many watts can it put out? If you are powering something small, then a low wattage inverter will suffice. If something large, then look for a 1000-watt or higher device – but here you really should try to know what you want to power first. Inergy, the manufacturer of my Apex, got a bit of a bloody nose when their unit, marketed to put out 1500-watts, was found that there was a 5-minute limit to that power level, after which it would kick back to about 550-watts. This was a surprise to many who were unhappy with it – but there was very good reasoning for it – it just wasn't realized by many buyers at first. Five minutes is long enough to run a hair dryer for a bit, heat something in the microwave, or perform a quick surge with appliances.

By limiting the output to 550-watts continuous output the battery life stretches considerably ensuring longer life of both the battery and the system. This has not been an issue for me, and Inergy's new Flex model increases their continuous energy out to 1500-watts due to customer feedback. And if you need more than that? There are manufacturers such as Goal Zero, Max Oak, and Titan that can put out 2500 to 3000 watts of continuous energy – but of course at a significant dollar cost. By not rapidly discharging and recharging the battery too often, the number of charge cycles can be greatly extended allowing an estimated 2000 or more recharges to my Apex insuring a long life.

Spec: Weight

The last specification is weight. Some small units such as Jackery's 120-watt model comes in at 6-lbs while their 500-watt is still a mere 13lbs. My Apex (1500-watts) comes in at 25lbs, while many others can weigh in much higher than this approaching 100-lbs.

Several models of Solar Generators for Comparison

Specification	Point Zero Titan	Maxoak AC200P	Ecoflow Delta 1300	Inergy Flex	Goal Zero 6000x	Jackery 2000 Watt	Jackery 1000 Watt	"Other" 1800	Rich Solar X1500	Bluetti AC50S
Battery Size (Watt-Hours)	2000	2000	1298	1069	6075	2060	1002	768	1500	500
Battery Type / Chemistry	Lithium	Lithium LiFeP04	Lithium	Lithium NMC	Lithium NMC	18650 Li-ion	18650 Li-ion	Lithium LiFeP04	Lithium Ion	Lithium Li-ion
Power Out Normal	3000w	2000w	1800w	1500w	2000w	2200w	1000w	1800w	900w	300w
Power Out Peak/Max	6000w	4000w	3300w	3000w	3500w	4400w	2000w	3048w	1000w	450w
Expandable Batteries?	Yes	No	No	Yes	Yes	No	No	No	No	No
Can units Daisy-chain	No	No	Yes	No	No	No	No	No	No	No
Sine – Pure or Modified	Pure	Pure	Pure	Pure	Pure	Pure	Pure	Pure	Pure	Pure
Charge Controller MPPT/PWM	MPPT	MPPT	MPPT	MPPT	MPPT	MPPT	MPPT	MPPT	MPPT	MPPT
Solar Input Capacity	1000w	700w	400w	400w	600w	800w	200w	400w	500w	120w
AC Input Capacity	1200w	800w	1200w	225w	600w	1164w	163w	480w	160w	90w
Weight (Lbs)	67	52.5	31	30	104	43.2	22	46	37.9	13.7
120v Outlets	6	6	6	6	2	4	3	2	2	2
USB-A Ports	4	4	4	2	2	2	2	4	4	4
USB-C Ports	2	1	2	2	2	1	1	0	1	1
12v Outlets	4	2	1	2	1	1	1	1	1	1
Price w/o Solar Panels	$3195	$1,999	$1,199	$1,350	$4,995	$2,099	$999	$2,497	$999	$369

Point Zero Energy

https://www.pointzeroenergy.com/

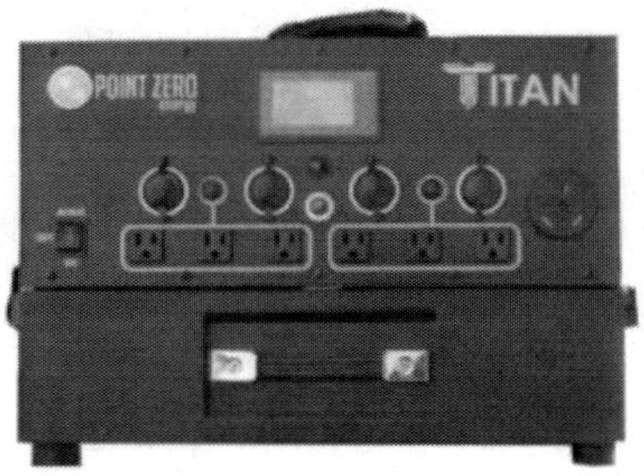

Point Zero Energy is the manufacturer of the Titan Generator – A heavy hitter in the solar generator market with key features including a 2500-watt sustained output, expandable batteries allowing more batteries to be added, up to 2000-watts charging from solar, 1450-watts from AC, pure sinewave inverter, and a 2000-watt standard lithium battery.

At the time of printing the Titan Solar Generator retails at $3195.

This solar generator is a workhorse and ideal for anyone needing large power capacities and the ability to expand with additional batteries. The high charging wattage rate would allow you to connect to a large solar array using the Titan as the power source for an off-grid cabin, RV or residence for extended periods of time. The weight of the unit is 32lbs plus 35lbs for each added lithium battery. Additional batteries run $1395 for 2,000-watts. For maximum charging you will need to purchase additional accessories.
Quick notes on the Titan Solar Generator:

Battery Capacity	2000-watts ($1.60/Watt)
Continuous Output	3000-watts
Expandable?	Yes – Multiple 2-kW Batteries can be added
Max Solar Input	1000-watts
AC Power input	1200-watts
Warranty	2-Years
Price	$3195
Notes	Check shipping availability for the Titan series. Short inventory supplies may exist making these hard to purchase.

Maxoak/Bluetti ACP200P

https://www.bluetti.com/

The Maxoak ACP200P is a relatively new power system by Maxoak which has a high output of 2000-wh using a 2000-wh lithium battery. Strengths of this power box include the ability to put out a higher wattage than many other devices and will work well to sustain light power loads overnight or even provide power for an RV (Less the AC). Drawbacks with the device include lack of battery expansion, however it does support 700w of solar input in addition to 400w of AC input combined allowing a full recharge in as little as under 3 hours. This unit has the power to keep a refrigerator or freezer running overnight. The unit retails at $1999 but has been on sale for $1699. If an expansion battery is not needed for your solution, then this should be a top consideration.

Quick notes on the ACP200P:

Battery Capacity	2000-watts ($0.90/Watt)
Continuous Output	2000-watts
Expandable?	No
Max Solar Input	700-watts
AC Power input	500-watts
Warranty	2-Years
Price	$1799
Notes	Good charge capabilities, capacity and output power makes this a good option where battery expansion is not necessary.

Quick notes on the AC50s:

Battery Capacity	500-watts ($0.86/Watt)
Continuous Output	300-watts
Expandable?	No
Max Solar Input	120-watts
AC Power input	120-watts
Warranty	2 Years
Price	$429
Notes	This is the smallest of our group. At only about 13 and ½ lbs, this unit is highly portable. It is a lower cost unit at only $429 with a corresponding lower charge rate at 120-watts and lower output at 300-watts. This would make a good light-weight and light-use solution.

Ecoflow Delta

https://ecoflow.com/

Ecoflow manufactures several solar generators – with the Ecoflow Delta 1300 being the top of their product line with a 1260-watt Lithium battery and being able to output 1800-watts of pure-sinewave power consistently. The unit supports up to 400-watts of input solar allowing it to recharge in up to 4 hours of time and boasts a recharge time of less than 2 hours when charging via AC power. The unit does support expanded batteries.

On the low-end they market a River 370 selling for $499 with a 412-wh lithium battery and a 300-watt pure-sinewave power out. At only 11lbs this is a small portable device that could be useful for small power needs or being able to power small medical devices such as a CPAP machine. The unit does support charging through AC, car charger or a single 100-watt solar panel with a charge time of between 6 and 7 hours.

Ecoflow has 7 products in their lineup as of early 2021 ranging in price from $499 up to $1249 for their Ecoflow Delta product. Later in 2021 they should be shipping out new models that may make the Ecoflow one of the top options on the market.

Quick notes on the Ecoflow Delta 1300:

Battery Capacity	1260-watts ($1.11/Watt)
Continuous Output	1800-watts
Expandable?	No (Newer Ecoflow Models coming out will support expansion)
Max Solar Input	400-watts
AC Power input	1200-watts (With Optional X-Stream Charger)
Warranty	2 Years
Price	$1399
Notes	Not the best of our bunch, but upcoming models from Ecoflow will offer much higher capacity and expansion.

Inergy Flex

https://www.InergyTek.com/

Inergy is a Solar Generator and Solar panel manufacturer which has just two a few products and accessories compatible with their solar generators. They have been around for a few years and are still relatively new, but they have been a leading innovator with other manufacturers following some of their design innovations. Their product has gone through three generations of development with the latest – the Inergy Flex now available for order at a current price of $1350. Their newest solar generator is a good medium size product which provides 1500-watts of sustained power from an 1069-wh Lithium battery. The Inergy Flex does check off many of the checkboxes I like on a solar generator – a 1500-watt pure sinewave output, an MPPT charge controller supporting up to 600-watts of input power from panels, relatively light weight at 30lbs making it portable, and the ability to expand with additional 1100-watt battery modules (up to 96) allowing powerful expansion if necessary. Though not shipping at the time of the writing of this book, I do have experience with their prior model, the Inergy Apex solar generator. Inergy uses a metal casing for their products with a good level of quality. The largest criticism of the Apex was the limit of only putting out 1500-watts for about 5 minutes, then reducing sustained output after at 550-watts. The newer Flex model eliminates this restriction allowing a 1500-watt sustained draw from the solar generator.

Quick notes on the Inergy Flex:

Battery Capacity	1,069-watts ($1.40/Watt)
Continuous Output	1,500-watts
Expandable?	Yes – 1069 Expansion Batteries, 96 Max
Max Solar Input	400 watts (Optional Charging module will increase this amount)
AC Power input	200-watts with optional Quick Wall charger
Warranty	2 Years
Price	$1500
Notes	Great middle-of-the road unit, relatively light at 30lbs. Biggest issue has been availability due to component parts with orders sometimes extending over a year until delivery.

Goal Zero Yeti 6000x

https://www.goalzero.com/

Goal Zero was one of the first companies in the Solar Generator business and continues to be a leader. Today their product line tops off with the Yeti 6000x Portable Power station coming in at a whopping $4995 but boasting some of the highest specifications in the market. The unit comes with a 6071-wh battery capable of sustaining 2000-watts out with pure-sinewave AC power. Goal Zero is often weak in some areas with their products and that is the case here with 600-watts of solar charging capability, but it does have 600-watts of AC charging ability. This means that as the most expensive of the heavy hitters it does have the best ability to charge, but for a temporary backup solution this may still be a winner. The 6000x also weighs in (literally) at 106lbs making it less portable than many others.

Goal Zero does win with battery expansion and the ability to tie the unit in with the home using battery expansion kits and even home power cutover accessories.

Goal Zero also has many smaller power solutions. They also are the only of the solar generators listed here that can be found in some retail stores, including Bass Pro shops, though not all models are carried.

Quick notes on the Goal Zero 6000x:

Battery Capacity	6071-wh ($0.82/Watt)
Continuous Output	2000-watt AC
Expandable?	No
Max Solar Input	600-watts
AC Power input	600-watts
Warranty	2-Years
Price	$4999
Notes	Weakness in the use of proprietary panels. Battery is also rated at only 500 charge cycles. High battery capacity makes this viable for whole-home use through interfacing through home circuit panel.

Jackery 2000 & 1500 Power Solutions

https://www.jackery.com/

Jackery manufactures several different models from a small 160-watt solar generator up to their newest Jackery Explorer 2000 – a 2200w AC system with the ability to recharge with 800-watts of solar input or through the integrated AC Charger. This newest Jackery is one of the first that allows a rapid recharging of the unit to allow the generator to be used over a period of days if needed.

Their 1500-watt Power Solution also offers a rapid charging capability making both the 1500 and 2000 versatile for use during emergencies that may last days or longer. Aside from these two larger models, Jackery also has several models in the 1000, 500, 300, 240 and 160-watt size ranges though these smaller units all have much slower charging times required making them great as convenience tools but less useful during a long-lasting power event where recharging is important. None of the Jackery solar power systems support external battery expansion. Jackery is a company that has been around for a bit with several models that are available along with solar panels and complete power solutions. For emergency power, I would recommend their larger units without question. Even without expansion batteries, the newer larger recharge capacity makes these better for use when needed. Recharge from AC power, or recharge from Solar. With either method you can be fully recharged within a very few hours for continued use.

When planning to use Solar panels for the Jackery generators, these generators do not use the MC4 connectors adopted by most panel makers. The Jackery panels are much more expensive than most third-party panels are. You can use third party panels, but you will need adapters.

Quick notes on the Jackery 2000:

Battery Capacity	2060-watts ($1.02/Watt)
Continuous Output	2000-watts
Expandable?	No
Max Solar Input	800-watts (Jackery SolarSaga Panels)
AC Power input	800-watts (Built-in Power Cable)
Warranty	2-Years
Price	$2099
Notes	Weakness in the panels – Requires their own panel using Anderson connectors.

UNNAMED "1800"

Due to the extremely high cost per Watt for this unit we do not recommend it. We are not providing the manufacturer's name – however this is a unit currently sold and in our opinion is an extremely overpriced solution. This model solar generator is being included because it is being marketed towards the prepping and readiness community. I won't take too much time on reviewing the specifications of it – however just looking over the size of the battery at a mere 768-watt/hours along with the price makes this the most expensive in our list per watt/hour. At over $3.25/watt based on price to stored battery capacity. A decent solar and AC power input will allow the unit to fully recharge rapidly, but mostly because of the small battery size. My recommendation would be to look at other options before considering this as a power source solution.

Quick notes on the UNNAMED "1800":

Battery Capacity	768-watts ($3.25/Watt)
Continuous Output	1800-watts
Expandable?	No
Max Solar Input	Not Specified
AC Power input	Not Specified
Warranty	Not Specified
Price	$2497
Notes	The most expensive product in our comparison at a Stored Watt/Dollar estimate. No expansion, no specified charge rates. No warranty specified on their website.

Rich Solar X1500

https://richsolar.com/

The Rich Solar X1500 came onto our radar recently before working on this book. This unit is the same, based on specifications and the photos, as the Bluetti EB240 which comes in at almost twice the price. The specifications of the Rich Solar X1500 place the unit in the middle-of-the-road for power solutions in our chart, putting out only 1000-watts of power but from a beefy 1500Wh battery. At a price of only $999 at the time of this writing, this unit should be a good value for someone looking for an emergency or backup solution with a large battery.

One of the drawbacks of the X1500 is the limit of only about 260-watts of AC charging input, but it does support up to 500-watts of solar input allowing a relatively rapid recharge from solar. The X1500 is a great middle-of-the-road solution at a good price.

Quick notes on the Rich Solar X1500:

Battery Capacity	1500-watts ($0.93/Watt / $0.67/Watt based on discounted price)
Continuous Output	1000-watts
Expandable?	No
Max Solar Input	200-watts
AC Power input	200-watts
Warranty	24-Months
Price	$1399/$999 (Check Website)
Notes	This unit has weak solar and AC Charging, and no expansion. However, for the discounted price under $1000 and capacity of 1500-watts batter, 1000-watts output it is a very price attractive unit – At their discount price of $999 this is the cheapest per watt of the unites we have reviewed. (Check their website)

Comments about some of these brands

Before moving to the next topic, I do want to mention one thing about some of the manufacturers I just covered in the prior few pages. Solar Generators are still relatively new products. Some of these manufacturers are still new to the market and have some excellent products. The combination of a smaller company and an innovative product can sometimes lead to lack of inventory and availability. Some of these companies have sold out of their products during a given year leading to several months waiting for a product. This has happened to me leading to months of wait times – though I strongly like their products and am a loyal owner, I eventually cancelled my Inergy expansion battery after 15-months due to the need for a working solution and no longer being able to wait.

Goal Zero – though in some ways not as innovative is a readily available product. As is Jackery and Ryobi. When deciding on a Power Station or Solar Generator for purchase know that some may take weeks or months for delivery while others may be readily For this reason, I recommend you pay close review to our later chapter seven where we will discuss the actual building of two small solar generators. Using some of the ideas we present in that chapter could lead you to building a more cost-effective solution and possible a more flexible solution.

RV Integration

The integration of any of the power sources in with an RV is a perfect example of a small-scale off-grid power solution integrated into a home power system. Solar is so popular among RV owners that there are plethora of videos and "How To" books introducing solar to owners and showing how to best use it. Depending on where you live and how often you RV, many full-time RVers are now considering some level of solar installation to be common sense to run at least minor appliances fully from solar. Solar power in an RV in fact makes so much sense that a small roof installation of just 500 or so watts along with a moderate battery bank of 2-4 batteries can run most appliances and power needs (Except for that dreaded Air Conditioner) exclusively. With 2200-watts of Lithium batteries and my solar generator I have enough power to keep the lights and TV running, the refrigerator cooling through the night – though the Air Conditioner is still not possible without several thousands of watts of battery power – usually too much for most cases.

Problems with Solar compared with Fuel-based generators

Just like there are certain negative aspects of fuel generator systems, likewise there are several negative aspects of a Solar based solar solution. You need to weigh these factors when deciding to purchase or build your own solution if you decide to make solar a part of your power plan.

COST

The most significant point will be cost. The cost of a portable solar generator purchased off the shelf will be roughly $1.00 to $1.25 per watt hour based on battery size – not on power wattage output. So getting into a larger solar generator that will power some lighting, some refrigeration, and maybe some other devices overnight during a blackout will be in the $1,000 to $2,000 range vs. a fuel generator that can cost as little as under a couple of hundred dollars. DIY solutions will save you money. And if you have access to good batteries for a short-term solution you can save much more from Lithium.

Size and Capacity
The one that is often most confusing. Size and Capacity. A 1500-watt Solar generator that can put out 1500-watts continuously, assuming it also has a 1500-watt battery, can last for only an hour. My Apex is a 1500-watt inverter but with an 1069-watt battery – so at a continuous 1500-watts I would use up my battery in about 45 minutes. Then I am dead until I recharge. If I do this in the first hour of nightfall, I have no power until I can recharge in the morning with solar panels. But if I have a 1500-watt fuel generator, I can produce 1500-watts of power EVERY HOUR during the night. In a 12-hour night that is about 18,000 watts (And about 2-3 gallons of gas).

And again, the benefits…

- Silent operation – You won't attract attention
- Fumeless Operation – You can operate it indoors
- Light Weight – Especially with Lithium batteries, these can be very convenient to carry
- Free Fuel – Once you make the investment, you can recharge with solar panels. This is great if fuel cost or availability will be an issue.

4: Battery Generators Grown Up – Home Power & Rolling Blackouts

In the last chapter we opened the conversation of solar generators – portable battery powered systems which provide AC power through a power inverter allowing users to operate everything from a portable phone charger to a refrigerator, small AC unit, lights, televisions, or medical equipment when there is no traditional grid power available. The amount of power available from these units is limited – depending on how much of a load is added, and how long you are running it. My current primary solar generator is the Inergy Apex that I mentioned in the prior chapter, which provides me 1069-watts of stored power that can be used over several hours to keep my refrigerator powered, lights on, and a few necessities going.

But – overall the smaller solar generator systems we discussed in the prior chapter have limitations on power. They are convenient to carry and highly portable, and they carry enough power for a limited number of devices. Though they can be recharged from solar even while in use, limited power capacity does restrict these mostly to portable power needs and short-term power. Overall, I still recommend having a fuel generator available for longer term and higher-level power needs. But – There Is another great optlon that may work better for the home – a system that can provide power in your home during a blackout or during a rolling blackout. These larger systems will allow more power through the home and may provide the power needed to keep you up and running for those few hours until power comes back on.

Rolling Blackouts – A more Common occurrence in many areas

As I am writing this new chapter in the latter half of 2021 for the second edition of this book, we can look at the increasing number of rolling blackouts that are affecting more areas of the country. These rolling blackouts occur for multiple causes – extreme cold from winter storms, wildfires spreading through the northwest, excessive heatwaves, high winds such as in California, or any of several other causes. The fact is – our power infrastructure is weakening. The increasing difficulty in providing power when the system is stressed will continue to worsen, leading residents the need to take steps in ensure power reliability in their homes.

In whole-home power solutions, these solutions are often associated with Solar power. With whole-home solar solutions, there are two types – Grid tied and Off-Grid. Let's review these briefly before explaining large-battery power for the home for background.

Grid-Tied Solar Solutions

A grid-tied solar solution for the home provides your home the ability to generate power from solar panels, and that power is fed into a power inverter in the home and then can be used for powering appliances within your house. Surplus energy is fed back into the grid and to the power company, with a credit being issued to you (Usually). The catch with grid tied however is that when the grid goes down, so does the ability to harness any of the electricity you are generating from your solar panels. This is a simpler and less expensive solar power solution and does benefit the homeowner through reduced or eliminated electric bills. Credits received back for surplus power, if the electric company pays credits, are generally a fraction of what you get charged for power. Costs of an off-grid solar system will run in the tens of thousands of dollars.

Off-Grid Solar Solutions

An off-grid whole-home solar solution takes our power generation ability to the next level – in addition to generating power, we now store that power for use later. Surplus power will now be stored in batteries located in the home. The costs of this type of system jumps drastically due to the cost of the batteries, inverters and power infrastructure modifications made to the home. In addition to the thousands in cost for the solar infrastructure we are now adding additional costs of thousands to tens of thousands more in the battery/power storage costs.

Whole Home Power Storage

Our ability to store power for later use is not restricted to the off-grid solar solution we discussed above. Instead of installing batteries for use only with a solar panel system, we can just as easily add a battery storage system into the home and keep the batteries charged up through AC Grid charging. When we lose power during a blackout, our batteries will provide us power while the grid is out. Depending upon the size of our batteries, we can power the entire home including our AC or heating (Very expensive) or we can size our backup power to keep our necessities on for a few hours such as refrigeration, some lights, our security system, or anything we may deem necessary while power is out. If our concern is strictly for blackout protection, then recharging our batteries from the grid may be as far as we need to build our system. But we can go a step farther. We can install solar panels onto the home that will allow power generation while the grid remains down during an extended power interruption.

But – How much power do we need?

In a later chapter we will discuss how to perform a power audit. In the power audit you perform, you should consider only the necessities you need for keeping power up for a period of time. What appliances you need, and how much power you will need will be specific to your needs.

Possibly the most well-known home battery solution currently is the Tesla Power Wall. This device can be installed either inside the garage, or outside of the house. The Tesla Wall has become so popular, back orders through many of their resellers have gone beyond a year with Tesla now only allowing the sale of their power wall systems with the installation of their own solar panel solutions. This makes the Tesla Powerwall the most expensive of the solutions currently available.

Tesla Powerwall

Other manufacturers are on the market now with solutions that have similar capabilities from companies such as Generac, LG, Sonnen, Goal Zero, Sunpower, Enphase, Panasonic, Sol Donum and others. These systems all work similarly in that they can be interconnected into the home electrical panel and provide power throughout the house while the power is down. Some of these systems require installation from the manufacturers or resellers, while others can be installed yourself and integrated into the home power panel requiring an electrician for safety. These systems all have features that make them different from each other and careful examination should be made for a solution that best fits your specific need. For some a smaller system which can provide some power for the home may suffice. These, while smaller in battery capacity, offer some level of portability allowing use in a non-permanent installation.

Permanent Installation vs. Temporary and Portable

The next pages contain two tables of various power systems from multiple manufacturers. The first table primarily contains larger permanently installed solutions such as the Tesla Powerwall which are mounted in the garage or the side of the home. The second table on the following page presents systems from manufacturers which are all portable in design – these systems allow you to remove the systems and take them with you in case you need to relocate. This would be beneficial if evacuating for a disaster, or if needing to relocate with friends or family. Doing so may not be as quickly as grabbing a suitcase however – the Goal Zero for instance weighs in at over 100 lbs., with both the Titan and Vulcan being over 50 lbs. with battery. But – with just a few minutes it certainly is possible.

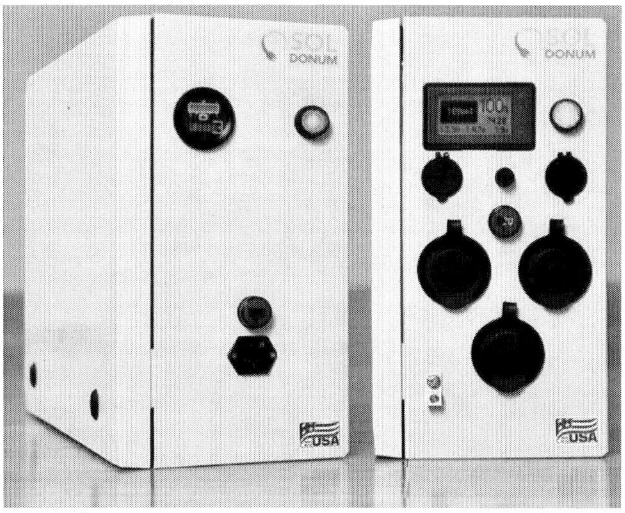

Sol Donum Vulcan Energy Store & Inverter-Charger

Title	Tesla Powerwall 2	SONNEN SonnenCore	LG RESu10	Generac PWRCell (3 Modules)	SOL DONUM Vulcan (5-kW)
Battery Size kWh	13.5 kWh	10-kWh	9.3kWh	9 kWh	Main unit 1344-wh Battery, Expansion with 2.6 or 3.6 kWh, unlimited units
Output Running	On grid: 7.6 kW full sun/5.8 kW no sun, Off-grid: 9.6 kW full sun/7 kW no sun	4.8 kW	5kW	3.4 kW	2 kWh (Two main units can be combined for 4-kW, higher capacity with custom solutions
Output Peak	9.6 kW	8.6 kW	7kW for 10 sec.	4.5 kW	4 kWh (Two main units combined for 8-kWh)
Battery Type	Lithium Ion	LiFeP04	NMC	Lithium Ion	LiFeP04
Install Type	Permanent	Permanent	Permanent	Permanent	Temp/Portable
Expandable Batteries	Yes	No	Yes	Yes	Yes
Weight	343 lbs	348 lbs.	165 lbs	280 lbs (W Batteries)	48 lbs Main unit, 45-52lbs for Battery
Operating Temp Range	–20°C to 50°C (–4°F to 122°F)	41-122°F	14 to 113°F	41F to 113F (5C to 45C)	-20°C to 50°C (-4°F to 122°F)
Rating	IP67 (Battery & Power Electronics) IP56 (Wiring Compartment	NEMA 12 (indoor)	IP-55	NEMA 3R/NEMA 6P	IP-54
Shipping	A Year or more delay in delivery	Yes	Yes	Yes	Yes
Battery Expansion Notes	Supports up to 10 AC-coupled Powerwalls	No Expansion, however, SONNEN does sell other models that are expandable.	Two units installed in parallel max.	Allows multiple configurations with up to (6) 3-kWh battery modules, Minimum of (3) modules to start.	The Vulcan system can be expanded with an unspecified number of expansion battery systems. Battery expansion units are available in 2.6 kWh and 3.6 kWh configurations, no maximum
Size (W Battery)	62.8 in x 29.7 in x 6.3 in	27w x 73h x 11d	8.9L x 17.8w x 19.0h	22" x 10" x 68"	15"H x 23" D x 9.5"W
Connectivity	Ethernet / Wi-Fi / Cellular (3G, LTE/4G)	Unknown	0	Rebus DC Nanogrid	Bluetooth for MPPT, Inverter
Environment	Indoor or Outdoor Rated	Indoor	0	Indoor or Outdoor Rated	Indoor/outdoor, Cold Weather Options (<32 degrees)
Warranty	10 Years, Batteries will have 70% Capacity	10,000 cycles or 10 years	10 Years – 60% of nominal energy for 10 years or 3180 cycles	10 Years	10-year limited warranty retaining energy capacity 80% at 10 years for up to 5000 cycles.
EMP Protection	No	No	No	No	Option
Retail	$10,500 (Before Installation) (	$9,500 (Before Installation)	Between $9,500 and $13,000	From about $7500 for 3kWh	$4,890 5-kW (Non-Solar Model)
Custom Configurations	NO	NO	NO	NO	YES

Table 4-1: Larger Battery Solutions for the Home

Title	Point Zero Titan	Goal Zero 6000x	SOL DONUM Vulcan (5-kW Configuration)
Battery Size kWh	2 kW	6071Wh	Main unit 1.4-kWh Battery, Expansion units either 2.6-kWh or 3.6-kWh
Output Running	3-kW	2-kW	2 kW (Two main units can be combined for 4-kW)
Output Peak	6-kW Watts	3.5-kWs	4 kWh (Two main units can be combined for 8-kWh)
Battery Type	NMC (LiFeP04 Option later)	Li-ion NMC	LiFeP04
Install Type	Temp/Portable	Temp/Portable	Temp/Portable
Expandable Batteries	Yes	No	Yes
Charging AC Size	400 Watts (Up to 1450 Watts with 2 optional 25A Chargers	600-watts	360-Watt for Inverter-Charger, 360-watts (600-watt charger available) for Battery Store
Charging PV Size	35v to 145v DC, 2000 Watts	400-watts	700-watt MPPT for Inverter-Charger unit. 700-watt MPPT in Energy Store Unit
240 Volt	No	No	Yes
Weight	67 lbs (32 lbs for Unit w/o Battery, Plus 35lb Battery)	106lbs	48 lbs Main unit, 45-52lbs for Battery Store
120v Outlets	6-15A	2	Main unit - (3) 15A 120v Ports
USB Ports	6 Smart USB, 2 USB-C	(1) 60-w USB-C plus more USB-A Ports	1-USB C1, 1-USB QC1
12v Ports	4 Regulated @ 13.8v (20A Each, max of 20A Total)	1	1-12v
Operating Temp Range	Unsure	32-104 F (0-40 C)	-20°C to 50°C (-4°F to 122°F)
Category	Med (2 kW - 5 kW)	Large (> 5 kW)	Large (> 5 kW)
Designed for	Portable Power	Home level backup power	Home level backup power
Shipping	Backorders often	Yes	Yes
Battery Expansion Notes	Stackable Batteries	No battery expansion options.	The Vulcan system can be expanded with an unspecified number of expansion battery systems. Battery expansion units are available in 2.6 kWh and 3.6 kWh configurations, no Maximum limit
Availability	Low	Yes	Shipping now
Car Charging	Unknown	12v/10A Car Charger (120-watts)	None
Size (W Battery)	18.5 x 12 x 12	15.3 x 10.1 x 17 in (38.9 x 25.7 x 43.2 cm)	15" H x 23" D x 9.5" W
MPPT Charger Notes	No MPPT Notes	0	The main unit has a MPPT Charge controller. Each battery expansion also has its own MPPT charger allowing for consistent overall charging as the system expands in size.
30A Outlet	Yes	No	Yes
AC Voltage In	120-v	120-v	120-v
Warranty	1 Year	2-Years	10-year limited warranty retaining energy capacity 80% at 10 years for up to 5000 cycles.
EMP Protection	No	No	Yes
Retail	$3195	$4995	$4,890 for 5-kWh capacity (Non-Solar Model)

Table 4-2: Smaller Battery Solutions for the Home – All portable models

SOL DONUM VULCAN SYSTEM

https://www.SolDonum.com/

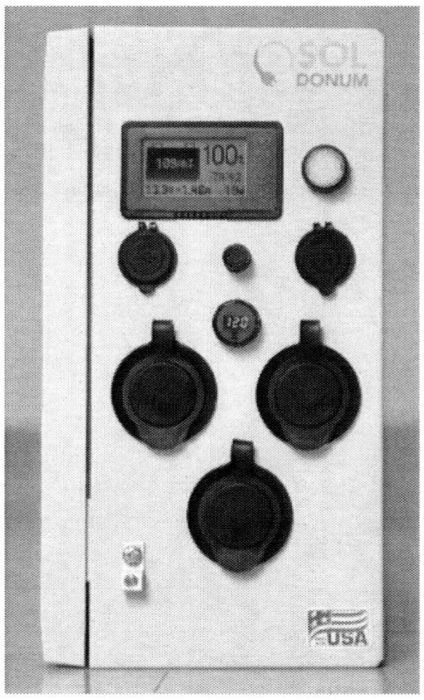

Sol Donum is one of the newest companies in our list with their new Vulcan system that is available as of 2021. This is the only one of the systems that we have included in both tables because it does fit in both the category of large and smaller portable systems. The Vulcan is not a wall-mount permanent system, but rather modular and portable. The base system itself contains 1.4 kW of battery, a 2000-watt Pure AC Sinewave inverter, MPPT Charge controller and AC Charge controller for grid charging. Alone the system price of $3495 is higher than any of our smaller solar generators with similar battery capacity, however when paired with their 3.6-kW Battery store system along with advanced features this unit makes it to the top of our list for multiple reasons. Some key features of the Sol Donum Vulcan are:

- Portable – The Inverter-Charger can be easily moved or used in any room of the house or taken with you in the need of having to relocate.
- Designed, Engineered and Manufactured in the USA with off-the-shelf components that can be easily replaced or switched out in the event of a problem making it highly serviceable.
- Expansion with additional "Energy Store" systems of 2.6-kWh or 3.6-kWh with no hard number of Energy Stores that can be added.
- Two Inverter-Charger units can be coupled together to provide up to 4-kW of power out.
- A 30A auto-cutover/passthrough capability allowing the Vulcan to provide "Hot" power without an automatic transfer switch or any intervention – This makes this a great solution for guaranteed power of medical equipment or refrigeration.
- These units can be installed and expanded without an electrician. If connecting into the house through the home electrical panel you still need an electrician who is licensed and knowledgeable about safe electrical power.
- EMI Filtered and EMP Shielding options (The only device listed in this chapter with EMP Shielding protection available.)
- Each unit (Inverter-Charger and Battery Stores) contains its own AC Charger AND MPPT Charge Controller for grid or solar panel charging. Simultaneous charging from AC and Solar panels also works for all units.
- Price for a 5-kWh Kit (1 Energy Store and 1 Inverter Charger): $6890.00 (Installable yourself if being used without home power integration into the breaker panel)
- Customizable Designs – Sol Donum will work with customers on custom solutions
- Field Replaceable Components – Depending on service needed, shipping back to Sol Donum may not be necessary!

Because of the versatility of the Sol Donum Vulcan system, we will be using this system in a later chapter with examples of adding a battery solution for brownout protections.

TESLA POWERWALL-2

https://www.tesla.com/

The Tesla Powerwall-2 as of the writing of this book is the current model available by Tesla. The Powerwall can store 13.5 kWh of power in its internal battery and putting out between 5.8kw to 7kw of power to your home. Additional battery capacity can be added through the addition of extra Powerwall systems. Charging is handled either through AC charging when the grid is up, or through Tesla's own solar panel system or through their solar roof system. Currently the cost of the Tesla Powerwall is among the highest of the Powerwall types of systems. Tesla, as of 2021, will only sell their Powerwall with packages including their own Solar panels or Solar Roof and will not be warranted or sold with the use of any other manufacturer's solar panel solution. This is one of the reasons that the Tesla solution is one of the higher priced solutions available.

Key features of the Tesla Powerwall systems include:

- Tesla is leader in Powerwall solutions
- 13.5 kWh Battery capacity, expandable by adding additional Wall units
- Output: On grid: 7.6 kW full sun/5.8 kW no sun, Off-grid: 9.6 kW full sun/7 kW no sun
- Wall mounted device, Indoor or Outdoor installation
- Retail: $10,500 plus installation (As of this writing the Tesla Powerwall is ONLY available with the purchase of a complete solution with Solar Panels or Solar Roof).

SONNEN SonnenCore

https://sonnenusa.com/

Sonnen is a US company headquartered out of Atlanta Georgia and produces a variety of whole-home solutions from their entry level SonnenCore model capable of putting out 4.8-kW of AC power out with a capacity of up to 10 kWh of energy storage. Their upper end home solutions include the ECOLINX and the ECO which are expandable in size and offer much greater capacities. The SonnenCore unit provides enough power to keep critical systems and lighting for your home through a rolling brownout or short-term power outage. Their entry unit starts at $9500 for the core unit, plus additional for installation. Adding solar panels for longer-term power outage and energy cost savings will bring the price up towards $15,000 or higher.

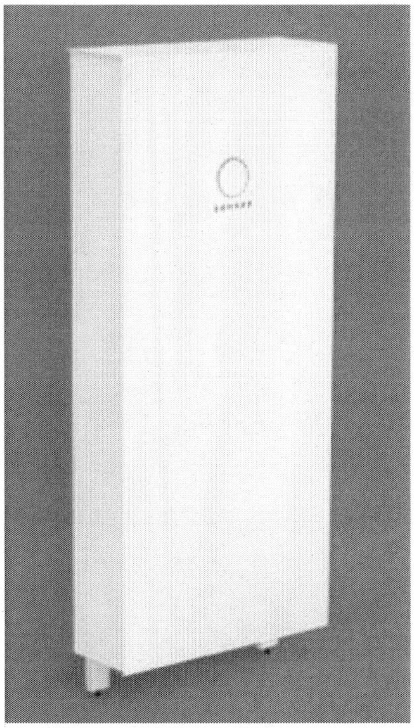

Sonnen SonnenCore Wall mount Solution

The SonnenCore unit is a single non-expandable wall mount unit. Warranty for the SonnenCore, like most similar products, is 10 years with a promise that the battery will still retain up to 70% of its original capacity at the end of ten years. The battery chemistry is Lithium Iron Phosphate which is one of the longer lasting and higher-capacity batteries available as well as being one of the safest chemistries available.

SONNEN SonnenCore Key features are:

- Battery Capacity of 10-kWh
- Standard power output of 4.8 kW, 8.6 kW Peak
- Permanent Wall installation (348 lbs.), Indoor only installation
- No Expansion, however, SONNEN does sell other models that are expandable.
- Warranty: 10,000 cycles or 10 years with reduced storage capacity at 10 years
- Retail price: $9,500 (Before Installation)

LG RESU10
https://www.lgessbattery.com/us/home-battery/intro.lg

LG Electronics is well known for a wide variety of home appliances and LG is also producing a series of home battery systems for backup power and off-grid power storage. The RESU series is a Powerwall type of power solution for the home with 9.3-kWh of power capacity expandable with up to two units in parallel.

The RESU10 is just one of several home battery systems, many of which are much more expandable to with additional batteries, and some of which are smaller such as the RESU3.3 with 3.3-kWh of energy capacity. Their largest capacity system is the M48128P6B with 4 Mega-Watts of storage capacity and is packaged as a 40-foot shipping container. Needless to say, the LG product line is advanced with a wide range of capacities and models.

LG RESU10 Home Battery

Key features of the LG RESU10 are:

- 9.3kWh Storage Capacity with 5kW continuous output
- 7kW for 10 sec.
- Permanent wall mounted unit weighing in at 165 lbs.
- Two units installed in parallel max.
- Warranty: 10 Years – 60% of nominal energy for 10 years or 3180 cycles
- Retail price Between $9,500 and $13,000

GENERAC PWRCELL
https://www.Generac.com/

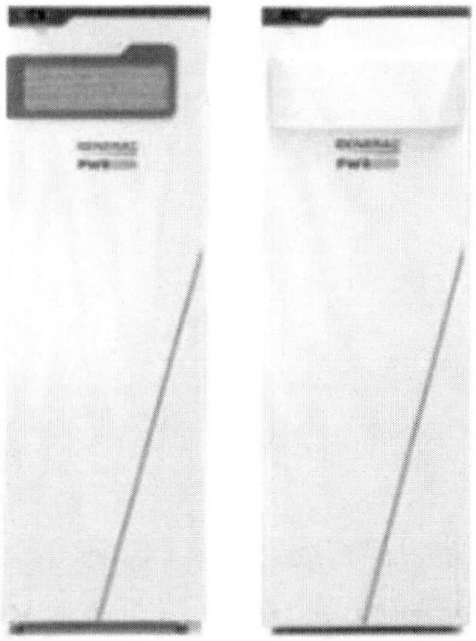
Outdoor Cabinet Indoor Cabinet

Generac, the leader of home off-grid generator systems in 2019 came out with their own Powerwall solution – the Generac PWRCELL. This system is a wall-mounted system that can provide home level power from an expandable home battery system. The PWRCELL is an expandable system with an indoor or outdoor wall mount cabinet housing up to 9 battery modules coupled with an Inverter for providing whole-home backup power.

Key features of the Generac PWRCELL include:

- 9 kWh battery storage capacity (With 3 battery modules)
- Power output of 3.4 kW up to 6.7-kW with 6 battery modules installed.
- Permanent wall installation with wall units for indoor or outdoor mounting, 280 lbs (With Batteries)
- Expandable batteries holding up to 9 Lithium-Ion batteries, Allows multiple configurations with up to (6) 3-kWh battery modules, Minimum of (3) modules to start.
- 10 Limited Years

POINT ZERO TITAN
https://www.PointZeroEnergy.com/

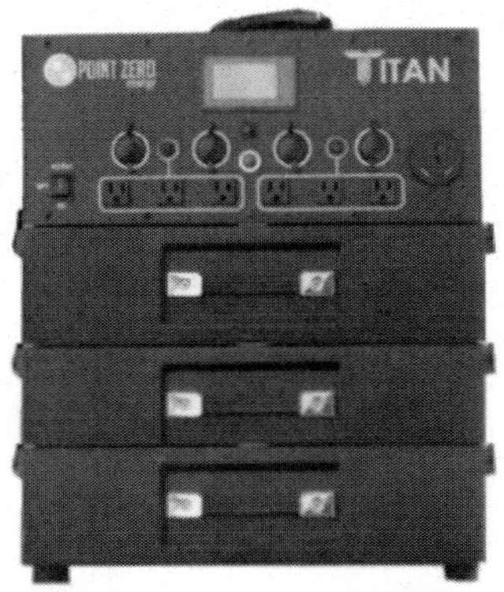

Titan Generator with 3 batteries

The Point Zero Titan is the largest of the Solar Generators that we had covered in the previous chapter. This unit is a large portable solar generator which supports battery expansion, 2-kW of capacity capable of putting out 3000 watts of power. The unit supports multiple 2000-watt batteries in a stacked configuration, a 30-Amp 120v RV outlet, an assortment of multiple 120-v, USB and 12-v DC Ports, and the ability to be powered into your home to provide 2000-watts of power into the home or providing power by directly connecting to the solar generator without tying it into your home power distribution.

If you need to have the Titan available to power equipment automatically during a brownout, then you will need to either have an automatic switch added to your home, or you may opt to run those devices directly from the Titan generator and allow the generator to simultaneously recharge. Unlike the Sol Donum Vulcan which can function by passing power through and auto-switching, this does impact operation on the generator by cycling the battery charging more frequently thus "Wearing" on the battery in a charge→discharge→charge→discharge cycle.

The Titan Generator has become a very popular system thanks to its high capabilities for a portable solar generator solution, however as a home solution it is not as large as our permanently mounted systems, but portability is a great advantage. This unit also has some of the best charging capabilities of all of the portable solar generators with being able to AC charge at 1000-watts, and solar charge at up to 2000-watts.

Unfortunately, the Titan has suffered shipping availability problems in the past and as of the time of this writing has a lead time of several months to being available.

GOAL ZERO 6000X
https://www.GoalZero.com/

The Goal Zero 6000x was profiled in our chapter covering portable solar generators, but the size of the unit easily allows use in this larger grouping for whole-home power. This is a portable device, like the Titan and the Sol Donum, not permanently mounted. At just over 100-lbs the 6000x is a portable system on wheels that can be moved and taken with you if the need to relocate were to arise. The 6000+ watts of energy stored is more than you will get with any other similar system on a cost per watt calculation, with 2000-watts maximum sustained output. One of the bigger drawbacks with the Goal Zero 6000x is the estimated recharge cycle count of just 500 recharges even though the unit has a Lithium Ion NMC battery.

Quick notes on the Goal Zero 6000x:

Battery Capacity	6071-watts
Continuous Output	2000-watts
Expandable?	No
Mounting Method	Portable unit (Not Wall Mounted)
Warranty	24-Months
Price Range	$4999
Notes	The Goal Zero 6000x is the second of our battery-based generators that crosses over to our large battery system list. Though not expandable, the 6071-battery capacity makes the Goal Zero large enough to for limited whole-home emergency power. The 500-cycle rating for the battery however does make this a less attractive solution for long-term use where frequent blackouts are of concern.

Reports of interest:

- The Most & Least Power Outages by State | MRO Electric

5: Project - Building a Home-Brew Solar/Power Generator

This is a project I first built myself prior to Hurricane Irma in 2017. Building a home-brew solar generator is an easy project to pull off with off-the-shelf components. In fact – Solar Panel/Charge controller kits are available at your local Harbor Freight stores called the "Thunderbolt 100-Watt Panel Kit" or online through Amazon, Home Depot, Walmart, or Lowes for ordering. Once you have a kit all you need to purchase is a Deep-cycle battery and AC Power inverter (Both available at Walmart and Harbor Freight for in-store purchases).

We are going to walk through the putting together of two solar power generators with parts that are readily available. The first will be a simple build – we are just going to lay out the components for use without worrying about mounting or putting them in a box. In the second project I am going to use similar components to build out a power solution using a large plastic ammo box capable of being a bit more portable use but a bit less flexible with our battery options. Using the basic design, you can expand on the system to add more solar panels, more battery storage, a better charge controller that accepts the energy from the panels, or a better battery inverter.

First – Let's review the parts we will be using:

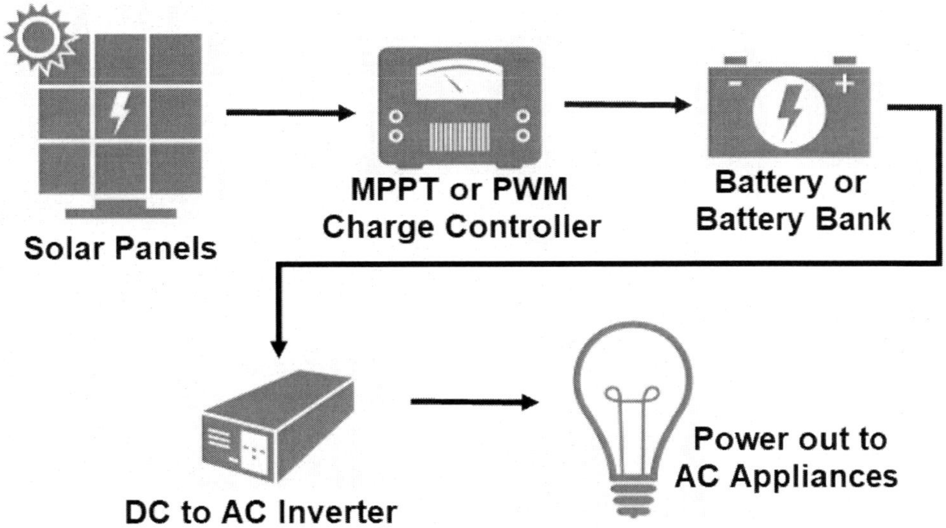

Solar Panels
One or more solar panels set up in Series or in Parallel from which we get our power that will be used to charge our batteries. We will build with one panel, but you can add more for better power input. Keep in mind that the efficiency of a 100-watt panel in great sunlight will be between 50% and 75% depending on the angle to the sun and the quality of the panels. So, you might want to

go with more than just one. The only way to guarantee 100-watts of actual power production is to use two panels.

Charge Controllers
Either a PWM Charge Controller (Less money, smaller loads, more basic design) or a MPPT Charge controller (More money, works with much larger loads, more advanced design, and more efficiency conversion of power into electricity. Many PWM charge controllers will be limited to 24 VOC of input power – Meaning that solar panels would have to be o greater than that voltage and could only be connected in parallel. I have a large 340-watt panel on my RV, with an VOC or over 50-volts meaning I could not use a PWM charge controller.

Battery or Batteries
You can have one or more batteries here to provide power. If multiple batteries are being used, they should be of the same voltage level in series or parallel depending on the design of your inverter and charge controller. More batteries will last longer and take longer to charge unless you increase your count of solar panels. Deep-cycle batteries are better than car batteries – and Lithium batteries are much better than Lead-acid batteries. If using multiple batteries in a battery bank, it would be best to use the same types.

DC to AC Power Inverter
This will convert our DC power from our batteries into our AC current used by our appliances. This includes TVs, Battery chargers, Refrigerator, CPAP machine, etc. You have a choice – Modified Sinewave inverter (Cheap) or Pure sinewave inverter (Best).

AC Battery Charger (Optional)
A wall charger can be used to keep your batteries topped off when the power grid is up or allow you to charge your batteries from a generator. This is generally not needed, but if you are building a solution for short-term power outages such as a rolling blackout, you could use an AC Charger rather than solar panels to keep your batteries charged. You need to match up the type of charger to the type of batteries you are using as voltage levels will vary between different chemistries of batteries.

Emergency Solar-Battery Power System – Design 1: Basic "Open" Design

Here are the components we need for this one:

1. Solar Panel (Can be purchased as part of a Harbor-Freight kit or a Solar Starter Kit, or as an individual solar panel). With panel selections – the kit will likely come with everything you need to get started. The Harbor Freight kits come with (4) individual 25-watt panels that connect for a total of 100-watts using SAE automotive style DC Connectors.

2. Harbor Freight (Or similar) Power Connection kit which includes SAE compatible connectors and an extension like those which already come with the Panel kit. Even if not using the HF

Solar kit, this connection kit is very inexpensive, plus it gives us connectors we may need to connect some devices to our battery (12-v Accessory/Cigarette lighter adapter).

SAE DC Power Connector, Unisex
Used by Harbor Freight Kits

MC-4 Solar Connectors, Male/Female
Used by most Solar Panels & Kits

3. PWM Charge Controller (Included in most start kits and in all HF kits). The HF PWM controller that comes with their kit works quite well – It also supports (2) 12-v LED Camp lights that come with the kit and are bright. These plug directly into the HF PWM Charge controller - about a $20 value if purchased separately.

4. A Deep-Cycle Battery (Marine Battery for instance). We can use a Car battery for our tests, but car batteries are not designed for the slow charge-discharge as our marine batteries and deep-cycle batteries are so will not hold up as well. You could optionally use a Lithium battery such as a LiFeP04 battery. If you do make sure you have an appropriate charge controller – The Harbor Freight charge controller is a Lead-Acid based charge controller. (Voltages are different)

5. DC to AC Power Inverter (A small one can be purchased for as little as $20 from Walmart, Harbor Freight, or most auto stores. A Pure-sinewave model for under $50 from Amazon). I have several that I have listed in chapter 8 showing the sinewave quality. In inexpensive inverter under $50 will power a laptop and/or small refrigerator.

6. You may need cables – Solar Extension cables for instance or wiring cable. Size of the cable is important as larger cables are necessary for carrying high current. I would recommend a gauge of between 8 and 12 in general. The smaller the gauge number (Larger cable) the more current the wire can carry. If you will be using a high wattage inverter or multiple batteries, then you should use a larger cable. For connecting batteries together in a battery-bank you may need to go to a larger cable (Smaller gauge number) anywhere from a 6-gauge wire down to a 4/0-gauge wire, depending on your design.

With this design we can lay it out on the garage floor, portable table, workbench, dining room table or anywhere convenient. We will just be setting out the components and wiring them together. The Solar panel does need good sunlight to work but can be extended outdoors with an extension cable using the proper connectors as required by the panels. Solar panels ARE waterproof and can be left out in the rain – But all other components are not and must be kept dry and out of the rain.

Solar Panel Kits:

Renogy's Solar Kit for under $200 – No stands or lights but easier to add panels to.

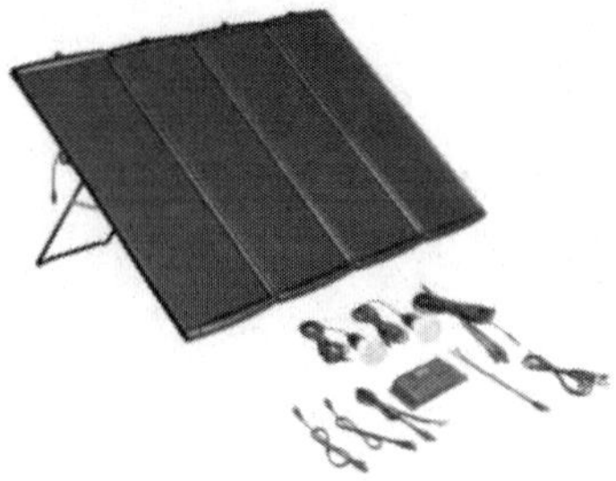

Harbor Freight's 100-watt Solar Kit includes Stands & Lights. Up to 4 kits can be connected.

SunGold Power's 130-watt foldable kit stores easy with a stand, retailing at $239 on Amazon.com

Warning: Working with Electricity can be Dangerous and even Deadly
In these projects we will be working with Electricity which can be dangerous in so many ways and can be deadly. Use extreme caution when working with any type of electrical projects and take all necessary safety measures. The 12-Volt electricity in our batteries is safer than the higher voltage AC electricity we will also be working with – however is still extremely dangerous. Shorting the battery terminals together can create a fire or rapid discharge of the battery leading to extreme heat and fire risk. Always take appropriate safety measures.

If you are not comfortable with working with electricity, then you should not work these projects unless you are properly supervised or assisted with someone with proper training and experience.

Step 1: Layout out the Battery, and Inverter

Here we are simply going to lay out these first two components. Even without the Solar Panel connected, you can use the Inverter with the battery to draw power in an emergency. The power inverter connects to the battery in this example through a set of jumper clamps due to the high current draw. Smaller power inverters will use a simple 12-volt cigarette lighter plug.

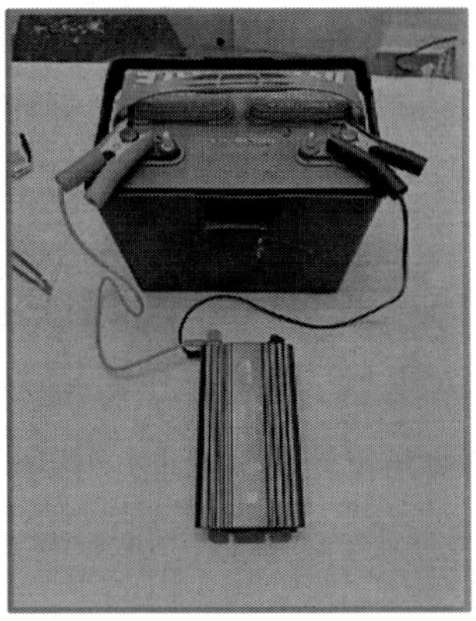

In the photo to the right, I have connected a 500-watt inverter directly to the battery. With just these two components we can power AC devices such as lights, small appliances, or televisions and radios, but without a charging method we will quickly drain down our battery. For small inverters, these often connect to the battery through a cigarette lighter connection. Larger inverters however pulling more power will pull more than 10 Amps which is usually the limit for Cigarette lighter connectors and cables.

Step 2: Add the Charge Controller to be able to recharge the battery

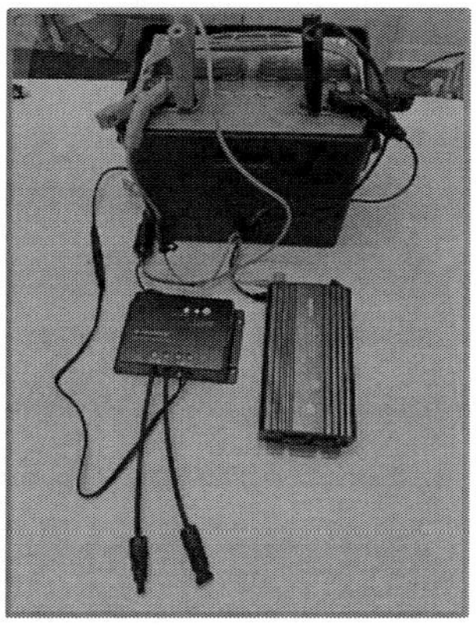

In this step we are adding the charge controller which will regulate the power coming in from the solar panels into the battery to keep it charged during the day. The charge controller is the device in the lower left of the image and has two short cords coming out using MC4 connectors to connect to the panels. If you are using the Harbor Freight panel, those panels connect with SAE connectors and the included connector cable for the charge controller to the panels will also have an SAE adapter on it. For general Solar use, the MC4 connector is preferred but is not always the choice of every panel manufacturer.

Step 3: Add the Solar Panel

In this step you will plug the solar panel in to the charge controller using cables that either came with the panel kit, or with cables you may have had to purchase separately. If you have purchased a panel starter kit, then the kit should have come with cables. Remember – the solar panel will be weather resistant and can remain in the rain. The other components cannot and must be protected from wet weather.

Most likely you will need at least a 20' extension to connect the panel to the PWM charger. Though the solar panel is designed to be in the weather and can be rained on the other components are not water resistant and must be isolated from the panel. The extension cable will accomplish that.

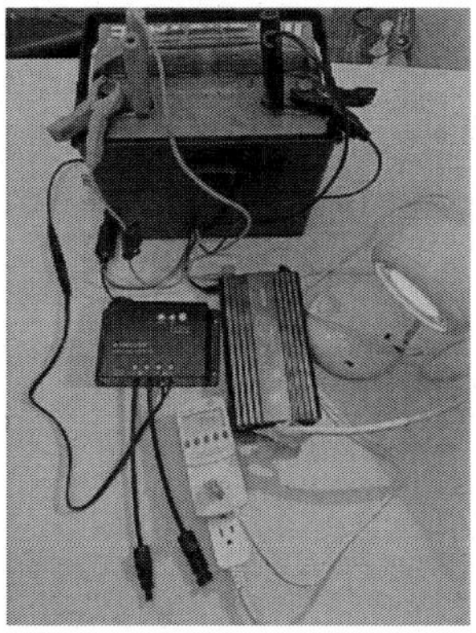

With the solution we have outlined in this section you will have a basic emergency solar panel system that can be used to power small AC appliances. If you were in a situation where you had no gasoline to power a fuel generator, or power appliances in your home this small generator would still provide you the power to recharge telephones, small radios, lighting (LED Lighting would be best), and even a two-way radio such as a ham radio.

Open Design Pros

The pros of the open design solution we have just provided can be summarized as follows:

- Flexible – Because we do not have our components pre-mounted and enclosed in a case, we can easily and quickly upgrade and swap out parts. Since we do not have an enclosed box that constrains our design, we can modify the design easily.
- Quick assembly – Parts can be laid out and quickly.

- Upgradable – You can easily add batteries, a better charge controller, more panels, or a larger or smaller inverter without usually having to change out other components.
- An open design can be grown – install your components on the wall of your garage and you can literally grow the system with expanding batteries or upgrading

The cons of the open design include:

- By not being in a self-contained box the open design is not as portable – multiple components must be laid back out when re-assembling.
- Not compact – You need more space to set the components up.
- Components must be selected and purchased individually.
- Requires more knowledge about each of the components and connecting them together.

Following is a Solar Generator less the panel that I have built, and I use in presentations:

Mounting it all Together

Once you get the parts together, you can get a assemble them using a section of plywood. I have built and mounted a solar generator here on a 2' x 2' board. A garage-based installation can be done just that simply. For this small project I am using a Renogy PWM charge controller which supports up to 25v coming in from solar panels and 30-amps, a small 460-Wh LiFeP04 battery (12.8v x 36ah), and a Bustek 500-watt Pure Sinewave inverter. I use this demo system as a working demonstration when I give classes. This unit is small, but has plenty of power to keep my refrigerator and freezer up and running long term without the grid running.

On the next page I am giving a breakdown of the components I am using as well as changes I would be making to upgrade this for more power. But even a small system like this one will provide power in the house, or even provide me some power in my RV.

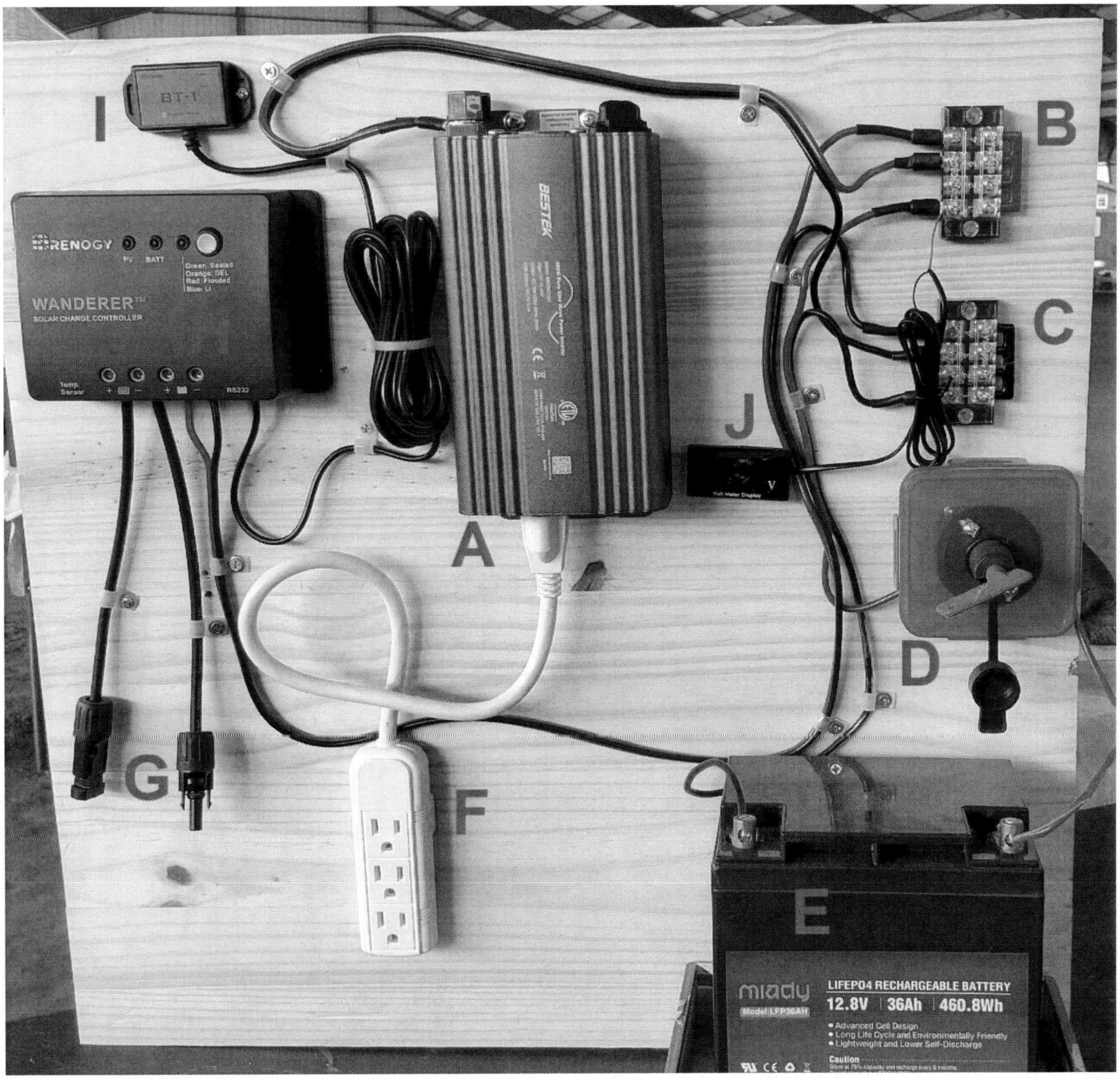

A) Bustek 500-watt Pure Sinewave Inverter. I selected a 500-watt Pure Sinewave inverter first because for light use 500-watts will allow me to power a few items including LED lighting, my refrigerator, and my freezer. A Pure sinewave is always my preference vs. a Modified sinewave is not a clean AC signal and more problematic. Also – A 500-watt inverter has less parasitic drain on the battery than a large12v r inverter would have. If installing into a Camper though, I would not add less than a 1000-watt pure sinewave inverter.

B) Positive 12-volt power distribution bus bar.

C) Negative 12-volt power distribution bus bar.

D) On/Off DC Power Switch (Installed for safety to disconnect the battery)

E) 12.8v LiFeP04 Lithium 36ah Battery (460-wh). For more power storage this battery could be upgraded to a larger battery. In my camper I have a single 1408-wh battery with three times the energy storage capacity. For home energy backup during a power outage I would recommend a larger battery than I have here to ensure longer operation.

F) Simple power strip to add additional 120v outlets.

G) MC-4 Connectors to connect to external solar panels. I have 25-foot cables with parallel connection capabilities to attach up to 300-watts of power to this model system.

H) Renogy 25-v/30-Amp PWM Charge Controller. For better performance an MPPT Charge controller could be added that would support higher voltage input and more current for higher power input. MPPT Controllers are also more efficient than a PWM controller and would be a better choice for a larger system. Also – The 25-volt limit restricts our solar panel configuration to be panels in Parallel only as 100-watt panels have typically a 18-voc rating.

I) Renogy Bluetooth module – Allows monitoring of power generation and battery from my cell phone.

Wiring Connections

For this project I connected all devices – Charge Controller, Inverter, and LIFeP04 battery to the positive and negative bus bars for convenience. Since the model I am building is just for demonstration and will not be powering anything large, I used 10-gauge Red/Black zip power wire throughout the build. In a home-use or RV-use configuration, my preference is to use no smaller than an 8-gauge wire for my solar panels to charge controller and from the charge controller to the batteries. For wiring from the Batteries to the Inverter, I would recommend a thicker and heavier gauge wire to carry the higher current load.

Look towards the chart on the next page for recommendations on wire size based on how much current you will have going through the wires. For the 10-gauge wire I used I can have at least 15-amps of current going through the cable. For connecting an Inverter to the battery/battery bank, a 4-gauge cable would be much better allowing you to move 60-amps through the wire. When connecting batteries together in a battery bank, an even thicker wire would be recommended – such as a 2/0 wire (Not 2-gauge) allowing 190-amps to move across the wires.

American Wire Guage Conductor Size Table

||
||
V

AWG	Diameter [inches]	Diameter [mm]	Area [mm^2]	Resistance [Ohms / 1000 ft]	Resistance [Ohms / km]	Max Current [Amperes]	Max Frequency for 100% skin depth
0000 (4/0)	0.46	11.684	107	0.049	0.16072	302	125 Hz
000 (3/0)	0.4096	10.40384	85	0.0618	0.202704	239	160 Hz
00 (2/0)	0.3648	9.26592	67.4	0.0779	0.255512	190	200 Hz
0 (1/0)	0.3249	8.25246	53.5	0.0983	0.322424	150	250 Hz
1	0.2893	7.34822	42.4	0.1239	0.406392	119	325 Hz
2	0.2576	6.54304	33.6	0.1563	0.512664	94	410 Hz
3	0.2294	5.82676	26.7	0.197	0.64616	75	500 Hz
4	0.2043	5.18922	21.2	0.2485	0.81508	60	650 Hz
5	0.1819	4.62026	16.8	0.3133	1.027624	47	810 Hz
6	0.162	4.1148	13.3	0.3951	1.295928	37	1100 Hz
7	0.1443	3.66522	10.5	0.4982	1.634096	30	1300 Hz
8	0.1285	3.2639	8.37	0.6282	2.060496	24	1650 Hz
9	0.1144	2.90576	6.63	0.7921	2.598088	19	2050 Hz
10	0.1019	2.58826	5.26	0.9989	3.276392	15	2600 Hz
11	0.0907	2.30378	4.17	1.26	4.1328	12	3200 Hz
12	0.0808	2.05232	3.31	1.588	5.20864	9.3	4150 Hz
13	0.072	1.8288	2.62	2.003	6.56984	7.4	5300 Hz
14	0.0641	1.62814	2.08	2.525	8.282	5.9	6700 Hz

https://www.solaris-shop.com/content/American%20Wire%20Gauge%20Conductor%20Size%20Table.pdf

Emergency Solar-Battery Power System – Design 2: Enclosed Portable Solution

The second design we are going to review is based around the use of a large plastic Ammo box to contain our battery and components. This containment allows our power box to be highly portable even though it may be heavy. Containing the design in a single box also prevents problems that could occur from components spread out, and the ammo box is a smaller and compact design which will hold most of our components – Battery, PWM Charge controller, 12-volt and USB ports, and SAE ports for connecting to solar panels and to an AC charger.

Ammo Power Box

In this design we are going to be using an external AC Power inverter. This allows us some flexibility in the manufacture/model and use only when needed. As a relatively small design the inverter I will use will likely be a smaller one such as my 120-watt modified sinewave inverter or slightly larger 300-watt pure-sinewave inverter. With wither I will plug the inverter in through the cigarette lighter plug.

1. Our Power box – Outside view 1
Here we are looking at the power box from the outside. What we see at the top of the box are (2) mounted ports used for output power. One is a 12-volt cigarette-lighter/accessory plug, and the second is a dual-USB port. On the side here we also see an SAE port. This port is used by a 12-volt charger and connect directly to the battery to charge the battery from the wall outlet.

2. Power Box – Storage Compartment
In the model box we have it here with a top-side storage compartment. In this, we have added an additional cigarette lighter/accessory plug, a power on-off switch, and we have run an accessory cord with Anderson plugs on it from the power that can plug into a Ham Radio.

3. Storage Compartment – Switch and Socket
In this closeup we can see the toggle switch that I picked up at Home Depot that powers the device and sockets. I added a second accessory socket for being able to plug in a cigarette-lighter based 12-v battery meter. Monitoring the level of your battery is important to prevent damage to the battery.

5. Power Box – Inside, Batteries and PWM Charge Controller

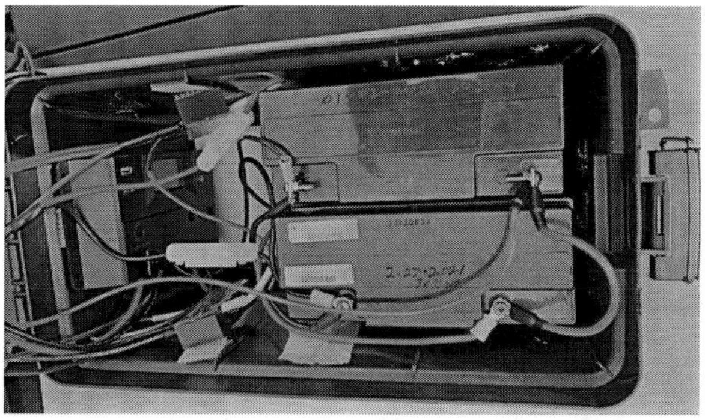

Here we see that I have added (2) 12-volt 20-amp/hour batteries which have been connected in parallel. This provides us a single 12-volt 40-amp/hour power source. Alternatively, you can use a single 35-Amp/Hour 12-volt battery – I just happened to have these two batteries to use which provided a bit more power capacity.

6. P**ower Box – Inside, PWM Charge Controller**

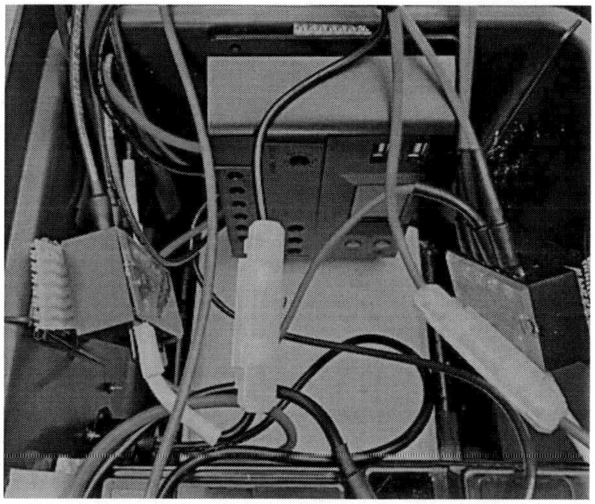

In this photo I am focusing in on the PWM Charge controller. Here I am using the PWM Charge controller which came with a Harbor Freight 100-watt panel kit. The charge controller is mounted against the back side of the box and fits perfectly. If you use another PWM Charge controller, most are of similar size. MPPT Charge controllers are generally larger more advanced devices and will not as easily fit.

Most production solar power boxes today use MPPT Controllers designed specifically for their custom design. I had both a Renogy PWM controller as well as this Harbor Freight Tools charge controller (With the solar kit, not the one they sell separately). I used this one for two reasons: (1) I just happened to have it and (2) is supports two LED Camping lights from the charge controller itself. This box is going to be small in design – 480-Watt/Hours, of which only half, or 240-watt/hours are usable. So this is a good but relatively small power box design.

7. Power Box Top – Ports Added

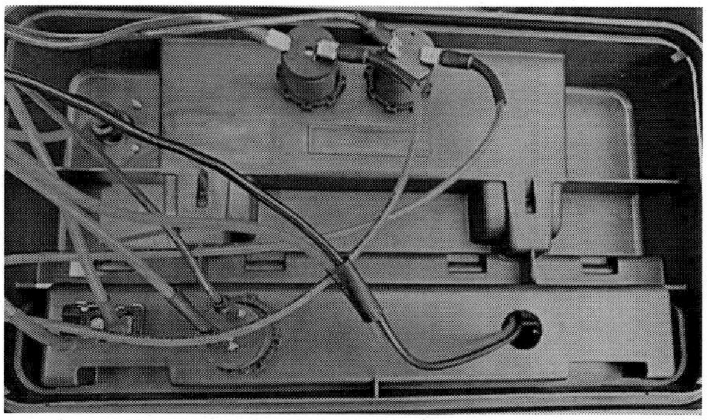

In this image to the right, we see the top of the ammo box where we have drilled out multiple holes and have inserted the cigarette lighter sockets, USB socket, and the SAE socket for the solar panels.

About the SAE ports

I used two separate SAE ports mounted from the side of the Ammo box and a second from the top. We needed to use two separate ports due to how the two charge sources connect for charging. The side port is connected directly to the battery and is used with an AC wall charger. The AC wall charging source needs to connect directly to the battery. The second SAE port connects from the solar panels to the PWM charge controller.

The solar panels I use for this box are the Harbor Freight panels that I have stored away. That kit came with camping lights and the PWM Charge controller that I used in the power box. The Harbor Freight panels use SAE connectors to connect, and to connect to the PWM charge controller. To support being able to use ANY standard 100-watt solar panels that typically use MC4 connectors I have an SAE to MC4 pigtail adapter (In the parts list on the next page).

What would I do in my next design?

From this design and build, I would make the following changes in the next box that I build:

1. Battery – The original box uses either a 35-amp hour/12-volt battery, or as in the box I built (2) 20-Amp hour/12-volt batteries. These are Lead-Acid batteries. With lead-acid you must stop use of the power box when the batteries reach 50% capacity, or 12-volts. Once you drop below 12-volts you begin damaging the batteries. In the next box I build I will move to a Lithium LiFePO4 battery.

   Miady is currently selling a 36-Amp hour/12.8v battery for about $130. This battery is lighter by over 50%, contains more usable energy per charge (Allowing closer to 100% discharge), contains a Battery Management System for preventing over charging or over discharging, and is the same physical size.

2. Toolbox for containing the power box – A larger toolbox allows more room for adding components. In the Ammo box design, there is very little room for a power inverter. Only a very small one (120-watt) would fit. In a larger toolbox we will have more room for adding a larger power inverter. I have two Bustek pure-sinewave inverters – either of which would fit in a larger box.

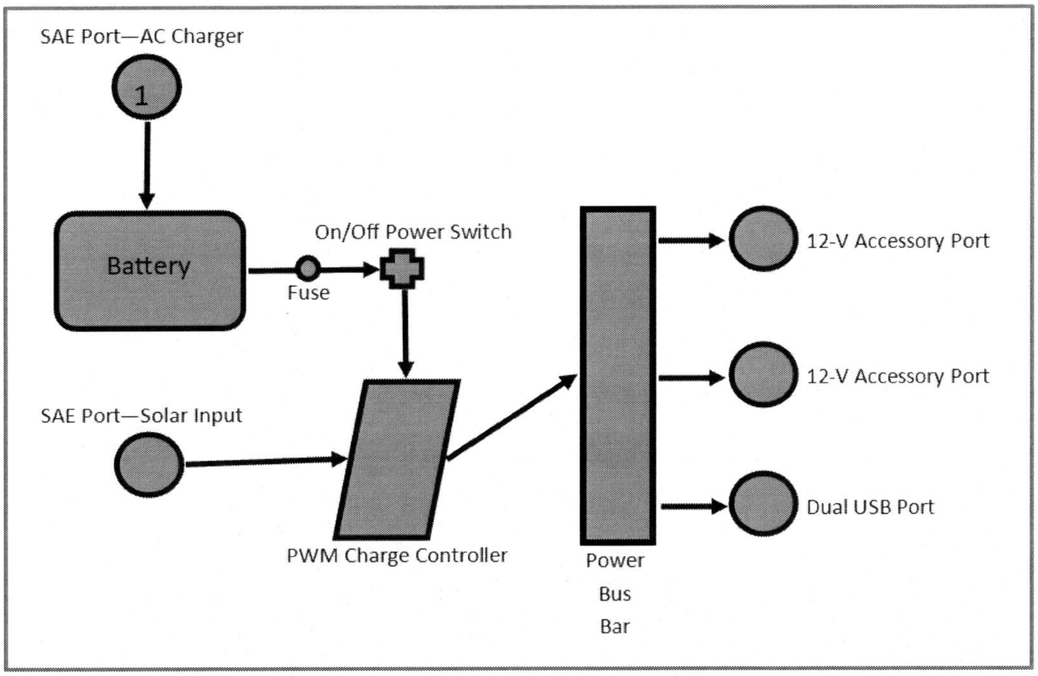

Layout of components used in the Ammo-Box Power Box

Parts List:

1. Large Ammo box from Harbor Freight (Or similar) https://www.harborfreight.com/tactical-ammoutility-box-64113.html
2. 35-Amp/Hour Battery from Harbor Freight or equivalent (I used (2) 20-Amp/Hour 12v Batteries)
3. SAE Connector, Electop Solar Weatherproof SAE Socket Sidewall Port https://www.amazon.com/dp/B07WCMX24L?psc=1&ref=ppx_yo2_dt_b_product_details
4. Solar Panel Connectors 10AWG Compatible with MC4 Connecter to SAE Adapter https://www.amazon.com/dp/B07MW1M1VB?psc=1&ref=ppx_yo2_dt_b_product_details
5. Harbor Freight 100-watt Solar Kit, PWM Charge Controller https://www.harborfreight.com/100-watt-solar-panel-kit-63585.html
6. 20-Amp Single-Pole/Single-Throw switch purchased from Home Depot https://www.homedepot.com/p/Gardner-Bender-20-Amp-Single-Pole-Toggle-Switch-1-Pack-GSW-121/100117400
7. (2) Terminal Switch Blocks/Bar Buses https://www.amazon.com/dp/B07Q81QSJJ?psc=1&ref=ppx_yo2_dt_b_product_details
8. (2) Cigarette Lighter Sockets https://www.amazon.com/dp/B07FVQ51CX?psc=1&ref=ppx_yo2_dt_b_product_details
9. Quick Charge 3.0 Dual USB Car Charger Socket 12V/24V 36W QC3.0 Dual USB Aluminum Socket Power Outlet with LED Voltmeter https://www.amazon.com/Charge-Charger-Aluminum-Voltmeter-Motorcycle/dp/B08F37WPDN/ref=sr_1_8?dchild=1&keywords=usb+socket&qid=1619300637&s=automotive&sr=1-8
10. Jebsens 4.8A 24W Dual USB Car Charger Volt Meter Car Battery Monitor

6: Getting it all Connected

With the use of a Portable generator, you need to decide how you are going to use the electricity and get that electrical power indoors. You have multiple methods of connecting your home to a fuel generator, ranging from simply running extension cords from the generator, to installation of a transfer switch mechanism that allows you to run power directly from the generator into your electrical panel for distribution.

Connecting the house – any way you look at it – is a task that must be planned out. If you go with the simpler extension cord method, realize that you first must have enough extension cords, and then realize that they must stretch into the house to provide power as needed.

SAFETY WARNING:
Any time you are powering your home from a fuel generator – Remember that the generator WILL produce harmful, deadly fumes. The generator MUST be located outdoors – Not in a garage. You MUST have proper ventilation for the generator to ensure that fumes do not move into the house where you and your family are.

Running a generator outdoors presents multiple issues:

- Safety – Keep the generator running several feet out from any door or window. Fumes can blow back into the house, garage, or even into the attic from the roof overhang if not moved away from the house.
- Security – If you are running a Fuel generator make sure you secure it. A Generator will be a target for thieves, as well as drawing attention to your home telling others you are prepared at some level. For this reason, I use a fuel generator during the day as well as a Solar generator/Battery Power at night.
- Be aware of dangers of electrical cords – If you are experiencing rain be aware that cords running through water are a deadly hazard. Cords can also be a tripping hazard.

To get the electrical power into your home you have several options:

1. **Extension Cords**
   This option is cheap – But it is also more difficult requiring multiple extension cords. That said, it may be your best option if your needs are rare. It does introduce hazards with energized cords running on the ground, but if your need is rare, this option does not require the installation of a permanent transfer mechanism to the home.

2. **Automatic Transfer Switch**
   The Automatic Transfer switch is the easiest functionally – but it is also the add cost and will be more common with larger stand-by generators or large solar systems to allow power to immediately be restored during a blackout. This system allows power to come from the generator directly into the home's electrical panel and then power only certain circuits within the house for emergency power. For most portable solutions, since the generators do not auto-start, an Automatic transfer switch is not necessary. Some generators however, if they have a battery start, can have automatic startup switches added allowing you to have a portable generator auto-start and provide power for the house through an automatic transfer switch. Note for this the generator needs to be housed outside in an enclosure where fumes will not get into the house.

3. **Manual Transfer Switch**
   The manual transfer switch allows power also to run into the home and through the existing home circuits but is a manual process to get power back up and running. If your backup power generator system is a manually started portable generator, then this system will work well as restoring power but will require the homeowner to take action to start the generator. This option is less expensive than the Automatic Transfer switch option above.

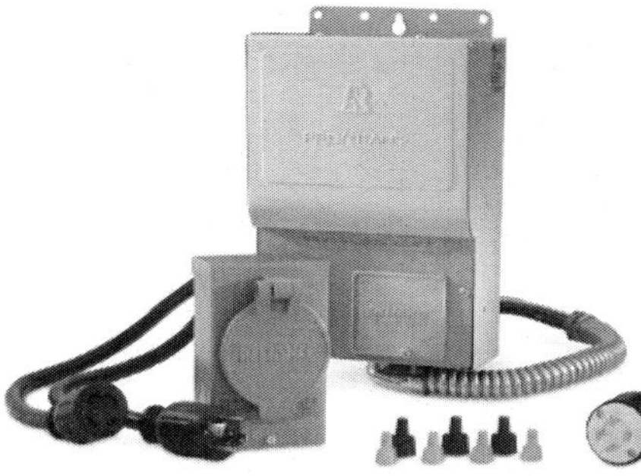

4. **Interlock Switch**
   This is the least expensive option which allows us to isolate our existing home circuit breaker from the normal incoming power lines, and instead power our entire circuit breaker from our generator. This choice still requires a manual intervention to re-energize the home, and it will also mean that the homeowner, when switching over to the generator in the process, needs to manually shut off circuits that are not needed or risk overloading the generator.

   For instance – If you are switching over to a 5,000-watt backup generator, in the process by turning all circuits not needed during this "Backup" mode such as you're Air Conditioner breaker or guest rooms, you can reduce the chances of overloading your generator. A large home Air Conditioner alone may take most or all your generator power which usually will trigger a shut down by most generators when overloaded.

   http://www.nooutage.com/interlock_kits.htm

There is another method which is NEVER recommended – And that is use of a Male-to-Male extension cord. These cords are also called "Suicide Cords". There are multiple reasons for this to be a hazard:

1. Once plugged into the generator, you have an unsafe second end to the generator with open contacts that are energized and live – This is an extreme shock risk!

2. Using this to feed power into your home will also back feed power to the power lines and create an electrical hazard for linemen working to restore power to the area.

3. This is ILLEGAL in most communities and you will open yourself up to criminal Negligence charges as well as liability if an accident occurs with one in use.

NEVER USE A DOUBLE-MALE EXTENSION PLUG TO BACK-FEED POWER INTO YOUR HOME. These are Dangerous and again – Illegal!

The installation of any of these transfer switch methods goes beyond the scope of this book and will be unique based on your specific needs as well as the switch method and manufacturer you are going to use. We strongly urge you to contact a licensed electrician only for proper installation based on your needs and your local electrical codes.

If you install a manual cutover switch, then getting power back to your house would generally follow the following steps:

1. Get out your fuel generator and prepare it for use. (Set it up outside only, 15 feet away from the home, with exhaust pointing away from your home, windows, and doors.)
2. Connect your generator to a pre-installed outlet that connects via the transfer switch to the home.
3. Start the generator
4. Switch your transfer switch from normal line grid power to your secondary generator source.
5. Selected circuits should now be active with power running to them.

Keep in mind with most manual switches, only certain circuits will be energized in the home based on a pre-planned strategy you will work out with the electrician installing the switch.

For an Interlock Switch, the cutover steps will be similar however there will be some differences since the entire breaker box will be energized rather than just those pre-planned.

1. Get out your fuel generator and prepare it for use. (Set it up outside only, 15 feet away from the home, with exhaust pointing away from your home, windows and doors.)
2. Connect your generator based on switch installation to the home.
3. Start the generator
4. On your Breaker box – AND THIS IS IMPORTANT – Switch OFF all circuits not needed during this blackout. Most likely this will include your Air Conditioner, Water Heater, Electric

Stove, and any rooms you determine not needed. FAILING THIS STEP MAY OVERLOAD YOUR GENERATOR.

5. Switch your interlock switch from normal line grid power to your secondary generator source.
6. Selected circuits should now be active with power running to them.

Specific circuits will now be energized based on your installation decisions. Only these rooms will have power available. Any lights and power outlets in these rooms will now work as normal using the power from the generator. When the power restores you will reverse these steps to switch back to the grid power.

Cost of a cutover switch generally will run from a couple hundred dollars to several hundred depending on what and how you are setting it up. Again, we emphasize, unless you have the training and experience, DO NOT DO THIS YOURSELF - Use a licensed electrician for all home electrical modifications. Electricity is deadly and you can easily make a mistake that could put your life or the life of another person at risk.

Transfer Switches – 120v and 240v

When planning a backup generator solution and planning to use a transfer switch or an Interlock, consider the voltage levels needed within your home. Many smaller generators will not provide 240-volt power for appliances requiring that much power, and likewise if you do need a generator and power for 240-volts, that needs to be planned also for your switch over installation. Several newer and larger inverter generators will now produce useful 240-volts if needed, though these may be limiting in current.

An RV 50-Amp system can be plugged into many 240-volt generators which can provide the split service that the RV is looking for, however usually you will NOT get the 50-Amps on each leg that the RV may be looking for. This does not mean it will not work – just that you won't get the full 50-amp service.

7: Performing an Energy Audit before you decide on a solution

In this chapter we are going to review the steps you will need to take to perform an Energy Audit. Whether you are planning a solution for a low cost and minimal power need for your home during a blackout, or you are looking to provide 100% power capability, or if you are looking for a solution for a Camper or off-grid cabin, the steps are the same. We need to look at what we decide are necessary items to have powered and operating throughout the loss of our grid.

To perform our audit, we are going to produce a list of appliances and devices that we want to have working during a blackout. For each item we will track the hourly wattage rating based on one of three sources:

1. The estimates from our Device List worksheet
2. The energy guide estimate that came with your appliance (If you have it)
3. The Wattage rating or Watt-hour's specification in the owner's manual for your appliance.

The estimates in our spreadsheet are just very rough estimates – and may be off by a great amount for your specific device. For instance, a new refrigerator that is energy efficient may only require 100-watts or less each hour, but an older less efficient model may require two to three times as much power. But – as an estimate these will give you a good start. The spreadsheet can be downloaded from our book website:

http://www.SmallPowerPlanning.com/EnergyAudit.xlsx.

Upon completing the spreadsheet, we will have an estimate of how much power will be needed. With this estimate we will be able to plan for how large and what type of power sources to look for. If your needs are minor, you may be able to get by with a small gasoline generator and/or a smaller solar generator. As your needs grow you can decide how you want to fill your power needs. Going with a larger gasoline generator may not be a big jump in cost, though going larger with a solar generator will get much more expensive quickly. To increase our capacity from 1600 watts to 3000 watts with a gas generator, I would need to invest under $500 for a larger model. To do the same with Solar, I need more solar panels, a larger inverter, and more battery capacity probably costing three to four times that amount.

Let's start off by looking at the spreadsheet and how we can use it to plan for our needs. The workbook contains four individual worksheets. Of these you will only need to update the AUDIT worksheet.

1. **<u>Tab 1: Instructions</u>** - This provides a summary of how to use the spreadsheet.

2. **<u>Tab 2: Devices List</u>** - This is a list of all of the items we have consolidated from several sources. Note that the power ratings for the items are *ESTIMATES* – You should check the exact power specs for your model of your equipment.

3. **<u>Tab 3: Audit</u>** - is the worksheet you will fill out. You can copy rows from the ALL ITEMS in Tab 2 (Devices List) or just those that you need into the Audit worksheet.

4. **<u>Tab 4: Sample Energy Audit</u>** - is a sample of our audit. Use this as a reference when performing your audit.

Part 1 of our Audit – Calculate power needs over a 24-hour window

PART 1: TOTALPOWER USED DURING THE DAY

Ref ID	Category	Sub Category	Item	Watts/Hr	Hours/Day Run	Watts/Day	On for Peak?	Peak Watts
14	AC/Cooling	AC	Window AC 5.000 BTU	600	8	4800	Yes	600
24	AC/Cooling	Fan	Pedestal Fan	60	8	480	Yes	60
25	AC/Cooling	Fan	Table Fan	75	8	600	Yes	75
33	Bathroom	Hair Dryer	Hair Dryer	1500	.1	150	No	
46	Computers	Laptop	Laptop	100	2	200	No	
47	Computers	Laptop	Tablet Computer	10	8	80	Yes	10
70	Electronics	Phones	Smart Phone – Recharge	6	6	36	Yes	6
102	Household	Garage Door	Garage Door Opener	350	.05	17.5	Yes	350
129	Kitchen	Coffee Maker	Coffee Maker	1750	.5	875	Yes	1750
155	Kitchen	Cooking	Oven: surface	1500	.5	750	Yes	1500
159	Kitchen	Cooking	Rice Cooker	800	.5	400	No	
160	Kitchen	Cooking	Slow Cooker	270	1.5	405	Yes	270
168	Kitchen	Freezer	Freezer – Chest – 15 cu. ft.	45	24	1080	Yes	45
177	Kitchen	Microwave	Microwave- 1500 Watt	1500	.25	375	Yes	1500
178	Kitchen	Refrigerator	ENERGY STAR Refrigerator (frost-free) 24 cu. ft.	75	24	1800	Yes	75
211	Lighting	Bulbs	CFL Bulb – 40 Watt Equivalent	11	16	176	Yes	11
212	Lighting	Bulbs	CFL Bulb – 60 Watt Equivalent	18	16	288	Yes	18
217	Lighting	Flourescent	Compact Fluorescent 25 Watt	28	8	224	Yes	28
244	Radios	CB Radio	CB Radio - Transmit @ 4 watts	6	2	12	Yes	6
251	Radios	Ham Radio	Ham Radio - Transmit @ 50 watts	75	1	75	No	
248	Radios	Clock Radio	Radio - Clock	10	24	240	Yes	10
263	Televisions	LED TV	32 Inch LED TV	60	8	480	Yes	60

	Total watts estimate for each Day:	**13,544**	Peak Wattage:	**6169**
	Total watts used each HOUR	**564**	(Hourly Basis)	
Estimated Overage % 15	Add Overage % for surges, unplanned growth	**15,575**	Peak Wattage:	**7094**
	Calculated Hourly use with 20% overage	**648**	(Hourly Basis)	

Figure 7-1 - Part 1 of our Audit spreadsheet calculating our 24-hour daily power needs

Using this worksheet, we can fill in rows and columns for the items we will be using during times when we have no grid power. Each row in the sheet is for a single item. If you have multiple lights, then you will need to add a single row for each one. Use the **Devices List** worksheet to copy an item that you wish to include in your audit into the **Audit** worksheet. Do this for each item to include adjusting the Wattage use for your specific models of equipment or use the estimated values we have provided.

The columns in the worksheet are:

1. Ref # – A Reference number for the item you are including in the audit.
2. Category – An overall category. The "AC/Cooling" category for instance has many types of items.
3. Sub-Category – A more specific category breakdown for the item being audited.
4. Item – A description of the item.
5. Watts/Hr – An estimated of wattage use per hour. Adjust this as needed for your equipment.
6. Hours/Day Run – Put in a number for the hours each day you will operate this item. A Refrigerator for instance will be set to "24" for 24/hours. A microwave may use ".25" for 15 minutes of use each day.
7. Watts/Day – Calculation based on hourly wattage use and the hours/day the item will be used.
8. On for Peak? – Set this to YES if this is an item you would be using at the same time as other items for a Maximum wattage use at any one time.
9. Peak Watts – A calculation of watts/hour for that item.

Once you add all the rows for the items you wish to account for the spreadsheet will calculate power totals. In our sample, as shown below here, we have a total of 13,544 watts being used each day. Below that number we have what our hourly average use is of 564 watts. For all the items we have set the PEAK parameter to YES for, we are showing that if all those items are active and running at the same time our Peak wattage will be 6,169 watts. This means if we want to run all those devices at the same time, we must have a generator source that puts out 6,169 watts continuously. This would rule out all of our portable solar generator systems. If we must have this amount of power need at any one time we will need either a very large battery solution such as a Tesla Powerwall, or a Fuel based generator.

		Total watts estimate for each Day:	**13,544**	Peak Wattage:	**6169**
		Total watts used each HOUR	**564**	(Hourly Basis)	
Estimated Overage %	15	Add Overage % for surges, unplanned growth	**15,575**	Peak Wattage:	**7094**
		Calculated Hourly use with overage	**648**	(Hourly Basis)	

Figure 2 - Our daily average of power use, peak wattage, and overage calculations

Below in the lower two lines of our worksheet, we have an "**Estimated Overage %**" value of 15. This overage amount accounts for either additional time you may need to operate devices or will consider power efficiency differences. Most likely the issue you may have is underestimating your power need, and this adjustment would help to account for that. In our example above with our 15% adjustment we will be using 15,575 watts/day of power, 648-watts hourly, with a peak wattage use of 7,094 watts. These would be the wattage estimates we will use when sizing a generator solution. This will also help us consider what we have identified in our usage tables that we might be able to drop if we need to for a temporary basis.

Our book and subject matter is focused on Small Power Planning after all – we are looking to develop solutions for short term and emergency use which will also be more cost effective. Adding a smaller 5000-watt portable gas generator for occasional use is much less expensive than large solar or large permanent whole-home fuel generators.

In our Audit worksheet we also have a second section which is headed "**Part 2- Overnight Power Needs**". This is on the same worksheet. The purpose of this second worksheet is to focus on a more limited power need at night – during a time when you may want to shut off the noisy gas or fuel generator and run off a silent power source such as a battery or solar generator, or during periods where you need to do maintenance on your fuel generator. If you plan to use a Solar generator overnight or battery powered solution, this second worksheet will be critical.

Hourly Wattage vs. Peak Wattage

Our hourly wattage numbers here are averages for the time. Our peak wattage is the maximum that we should be needing at any one time. After we cover Part 2 of this audit – planning for power overnight, we are going to delve more deeply into this. The reason is that knowing what your overnight power needs will be, we can plan around these numbers very carefully. Otherwise we could be without enough power to get through the night.

Part 2 of our Audit – How much power do we need overnight?

If your power solution is based only on a fuel-based generator then you can skip this part of our audit. I have included a "Night" version of the audit in case you are going to plan to use a solar generator or battery generator overnight. Because we are going to be more limited on how much power we can squeeze out of a battery-based generator solution, we need to look at our power needs differently. For myself, I will power off a noise-making fuel generator overnight and secure it to instead run off a battery-based generator. But – I must be much more careful on my power usage to avoid running out of power during the night. Look at

PART 2: OVERNIGHT POWER NEEDS (10 Hours)

Ref ID	Category	Sub Categry	Item	Watts/Hr	Hours/Day Run	Watts/Day	On for Peak?	Peak Watts
14	AC/Cooling	AC	Window AC 5.000 BTU	600	0	0	No	
24	AC/Cooling	Fan	Pedestal Fan	60	5	300	Yes	60
25	AC/Cooling	Fan	Table Fan	75	5	375	Yes	75
33	Bathroom	Hair Dryer	Hair Dryer	1500	0	0	No	
46	Computers	Laptop	Laptop	100	0	0	No	
47	Computers	Laptop	Tablet Computer	10	4	40	No	0
70	Electronics	Phones	Smart Phone – Recharge	6	4	24	Yes	6
102	Household	Garage Door	Garage Door Opener	350	.02	7	Yes	350
129	Kitchen	Coffee Maker	Coffee Maker	1750	0	0	No	0
155	Kitchen	Cooking	Oven: surface	1500	0	0	No	
159	Kitchen	Cooking	Rice Cooker	800	0	0	No	
160	Kitchen	Cooking	Slow Cooker	270	0	0	No	
168	Kitchen	Freezer	Freezer – Chest – 15 cu. ft.	45	10	450	Yes	45
177	Kitchen	Microwave	Microwave- 1500 Watt	1500	0	0	No	
178	Kitchen	Refrigerator	ENERGY STAR Refrigerator (frost-free) 24 cu. ft.	75	10	750	Yes	75
211	Lighting	Bulbs	CFL Bulb – 40 Watt Equivalent	11	10	110	Yes	11
212	Lighting	Bulbs	CFL Bulb – 60 Watt Equivalent	18	5	90	Yes	18
217	Lighting	Flourescent	Compact Fluorescent 25 Watt	28	5	140	Yes	28
244	Radios	CB Radio	CB Radio - Transmit @ 4 watts	6	2	12	Yes	6
251	Radios	Ham Radio	Ham Radio - Transmit @ 50 watts	75	.25	18.75	No	
248	Radios	Clock Radio	Radio - Clock	10	10	100	Yes	10
263	Televisions	LED TV	32 Inch LED TV	60	5	300	Yes	60

Figure 7-3 - Energy audit running over a limited time-period of 10 hours

Part 2 of our audit is similar to our first part – however we are only going to calculate for a few hours to cover our night times, and we will also zero out those items in our "Hours/Day Run" column that we will not be using at night. Here are some of the changes we are going to make:

1. We are going to calculate for a 10-hour overnight window of time. This will generally cover our hours when the sun is not up though that may vary depending on where you are. Adjust this as you need.

2. At night we are going to zero out our Window AC unit as we will assume that we won't need it at night to conserve power. We will zero out other items as well such as our Hair Dryer, Coffee Maker, Oven, Cookers and Microwave.

3. We will reduce the number of hours for many of our items. For our Refrigerator and Freezer however, we are going to run the full night so set the Hours/Day run cell for each item to 10 hours, or however many hours you need to carry through the night.

Now let's look at our summary for our night hours. Since we have turned off several devices for the night (No AC, not cooking, and other device reductions) we will notice that our peak wattage

		Total watts estimate for each Day:	**2,717**	Peak Wattage:	**549**
		Total watts used each HOUR	**113**	(Hourly Basis)	
Estimated Overage %	15	Add Overage % for surges, unplanned growth	**3,124**	Peak Wattage:	**631**
		Calculated Hourly use with overage	**130**	(Hourly Basis)	

hours and hourly wattage hours have both greatly reduced. We see a large difference between both numbers, and we need to delve more into why.

For PEAK we need to know what our numbers may jump to at any one time, so we are sure that we do not flip a breaker on our power source. In the example, I have included the garage door as a part of our peak, even though it will only run for about 15 seconds on open/close. And likely we won't open and close it more than two or three times, it must be taken into account for our peak power needs. Our garage door consumes 350-watts/hour, so a total of 1-minute, we are actually only using 350-watts/60 minutes, or 350/60 which equals 5.83 watts which we can round off to 6 watts. But – For those four 15-second periods our power source MUST put out enough to cover that 350-watts. (This would be an excellent use of our watt meter. Plug the garage door into the watt meter and check the readings AS you are opening and closing the door for an exact number.)

So, our peak number covers our max wattage rating that our power source must be able to support. As for our average number, this is the average power each hour we need to be able to support but is less than what our running power needs to account for and less than our peak power.

How does this play into our Planning?

Depending on if you are going to be using a Fuel power source or a Battery based power source, this is a huge difference.

If you are using a fuel power source – Let's say a small 2000-watt inverter generator for overnight use because it is quieter than our larger 7000-watt contractor generator – then our generator is capable of putting out 2000-watts *EVERY* hour during our 10-hour overnight window. With our fuel generator we can easily meet our PEAK wattage of 631-watts (Assuming we are basing on our overage estimate) every single hour. If we have fuel, we can meet this need. In a grid outage I keep at least (5) 5-gallon cans of fuel available always. So, I can run my larger generator for at least 2-5 days depending on loads and how much I am running the generators. Here again though we are focusing just on our 10-hour night window.

What if we are using a batter-based energy source. For instance – I have a Inergy APEX Solar generator. The generator as a battery that has about 1069-watts of energy available in the battery. It puts out a sustained 550-watts of energy, but it can surge to 1500-watts for up to five minutes. This meets our peak requirements – Since I will only be operating the 350-watt garage door for about 1 minute of time, I can meet the 631-watts of surge power. When not surging, my max power should be about 281-watts. (631 surge watts – 350 watts for the garage door). So, I can easily sustain 281 watts.

Now – Here is where our averages come into play. The averages are not as important with our fuel generator because all we had to worry about there was our peak power – Our generator can put out 2,000-watts per hour if we have fuel, so we can put out 20,000-watts of power during our 10-hour overnight window – a substantial amount. With those numbers in fact, we can splurge on our energy use – running the fans for the full ten hours and even running our small AC unit during the full night.

But – In this example we are using a battery-based source which has limited capacity. Our average overage power consumption here is 130-watts per hour. Over ten hours, this comes out to 1,300 watts of power we will need. But – We only have 1069-watts of power. So, unless we reduce our power consumption we will run out of power after about 8-hours. This might not be a problem – Our freezer and refrigerator are safe to go for 4-hours without power, so in the morning we simply restart our fuel generator and put the solar generator back to charging either from solar panels, or from an AC charging source.

Another option is to expand our battery – I have a 1.4-kWh LiFeP04 external battery that I can use as a charging source to my solar generator. Overnight I can plug the solar generator into the extra battery, and it will charge while in use, so by morning it will still be nearly fully charged. I will likely have to top it off and recharge the 1.4-kWh LiFeP04 battery, but I was able to easily make it through the night.

So – If we are using a Fuel generator, our hourly wattage estimate is not really of use – instead we just need to focus on our PEAK estimate. But if using a battery-based power source, being either a small solar generator like my Inergy Apex, or a larger one like a Tesla Powerwall or Sol Donum Vulcan we MUST take our hourly average into account. A Tesla Powerwall or Sol Donum Vulcan each store large amounts of power – easily meeting our basic power needs for 24-hours.

HOW WE NEED TO NOW USE OUR COMPLETED AUDIT

Now that we have completed our energy audit, it is time to start thinking about the solution you can purchase. To do this we now have even more considerations to consider:

1. **How much PEAK power do we need?**
   That is, the amount of power we need when the maximum number of devices we need active at one time are running – even if that means a 350-watt garage door opener running for 15 seconds. Without the ability to support our PEAK power needs we will crash our power source, either by flipping a breaker or the power source shutting down.

2. **How much power do we need over a set amount of time?**
   In our samples about we saw that a 1500-watt fuel generator would not only be able to power our needs for ten hours at night, but we could turn on extra items like fans, a portable AC unit, extra lights, and even more items. But my 1500-watt solar generator (With 1050-watts of battery capacity) would not get through the full ten hours much less run additional devices. And we stress – This is a number we need to know only if your power source is battery-based like a solar generator because these devices, operating at night, have a limited power capacity over the full night.

3. **What is our financial budget?**
   I can purchase a 1400-watt inverter gas generator from Harbor freight (1400-watts peak, 1100-watts continuous) for under $500 that will provide me 11,000 watts of power over a ten-hour night. But that same $500 will only purchase a 500-watt Solar generator that will provide me 500-watts of power – 1/22 of the power. That solar generator will run out of power in our example in less than 4 hours. Also – If you need Solar panels to recharge during the day, estimate about $100 per 100-watt panel extra (Not mounted). And some solar generators (Jackery) require either adapters to use with regular panels or require you to purchase theirs at a higher cost.

4. **Do you need clean energy at night?**
   Do you need your clean power source overnight such as to run indoors or to take to a shelter during an evacuation? This is impossible with a gas or fuel generator.

5. **Do we need the power source to feed into the house through breaker panel, or just to feed certain device through extension cords?**
   If you want to feed through the house estimate another $500 to $1000 for the equipment and electrician, maybe more. Also – If you want the switchover to be automatic, you may need a more expensive transfer switch depending on your backup power source.

To everyone I speak to about backup power I always recommend, if possible, going with both a fuel backup source (Gas generator) and at least a small battery-based system (Small soar generator). The battery-based system has so many advantages for short-term power and backup power indoors I would not be without one. But they can be expensive and do only provide a fraction of the power that you can get from a fuel-based generator.

My personal solution

The solution I use for myself is a Predator 3500-watt Gas-powered generator for my primary power, and for backup or running at night I use my Inergy Apex 550/1500-watt Solar Generator with an additional 2800-watts of Lithium LiFeP04 batteries which I assembled myself. I also own a 30' Travel Trailer, so if I am without power for days, my backup power sources will power both my camper (Including Refrigerator and AC) and allow me to run power into the house for my refrigerator and freezer, plus a couple of 10-watt LED lights in the house. I will be parked of course with the camper in front of the house and doing my actual sleeping and living in the camper.

The Predator 3500 can put out 3000-watts of power each hour and runs at least 8 hours on 2.5 Gallons of gas or longer. With enough fuel I can get 30,000-watts out over a 10-hour window.

The Inergy Apex stores 1069-watts of power with 550-watts of continuous output. At night with no solar available this equates to less than 2 hours of operation before being drained.

Notes about Efficiencies:
An Inverter Generator is much more fuel efficient at just 25% load meaning longer operation. This is because the generator throttles down to conserve power. A Non-inverter generator does NOT throttle down and will burn more fuel under a light load. Under a heavy load, both types of generators will burn fuel at the same rates.

Solar Generator Inverter Efficiencies:
This applies to all inverters, which a solar generator contains. When the inverter is on, even if no load is applied, there is a power cost in watts during idle use. The larger the inverter the greater the consumption usually and does vary depending on manufacturer/model. My Inergy Apex consumes 20-watts each hour when on whether a load is applied or not. Over ten hours this works out to 200-watts of energy wasted! You must take this into account when load planning.

8: Protecting against Rolling Blackouts

As our grid ages and continues to be put under more stress, as weather events and natural disasters continue to become more frequent and expected, and as our power infrastructure continues to lean more towards green energy solutions such as wind and solar power that are affected by even mild weather conditions, the need for us as homeowners will increase in the years to come to be ready for more power disruptions. Rolling blackouts are temporary power outages that last a few hours are becoming more common throughout much of the country. These are put in place by power companies to "Share the pain" of power loss by blacking out smaller portions of their coverage area for short periods of time – usually a few hours – and then restoring power to those areas while shutting down power to other areas.

In some parts of the country such as California, high wind periods force the shutdown of wind farms which the state is turning more to for power. During 2019 and 2021 Santa Anna Winds periods have triggered power outages to hundreds of thousands of residents which last not only hours, but in many cases days. The 2021 February Texas freeze is another example of power loss lasting from hours to days in length. And most recently in the Northwest United States where again wildfires have ravaged throughout multiple states and unusual levels of heat blackouts at times become normal affecting hundreds to thousands of residents.

In earlier chapters we have looked at several power solutions. We have looked at Portable fuel generators, small solar generators, and larger whole-home systems that can be integrated into the house. What are some good solutions for use if you have regular blackouts and regular rolling power outages?

Here are some things I would be looking at for protection during rolling blackouts:

- Power during shorter durations – Usually four to eight hours.
- Automatic startup when the power goes out – Important for refrigeration or medical needs.
- Ability to power at least critical appliances or circuits – Some lighting, Refrigeration, some other outlets. This may not include power for everything but will power enough to keep from losing food or critical medical equipment.
- Ability to grow or expand size based on need.
- Portability – Being able to take it with you during a evacuation, or if I relocate. A non-permanent installation is also better for renters, or occupants of condos, apartments, and townhomes.
- Possibility (But not requirement) for energy storage that can be used at more expensive peak times. (Charge during off-peak hours, providing power during peak hours).

- Battery based systems can recharge from the grid directly – no need to have solar panels. When power returns the system can simply charge back up and be ready for the next rolling blackout. (Solar Panels installation would be difficult if not impossible for renters and apartment dwellers.)
- If battery based, can the system pass through grid power when the grid is running, or does operation require pulling power through the batteries. If the system can pass through the grid power when the grid is operational, then the batteries will be kept in a topped off stage and only used when needed.
- Portable Generator based solutions can be modified to have an automatic start switch added if they have an electric start system. These are mostly DIY solutions and may require the assistance of a Mechanic to turn on. See the list at the end of this chapter titled "Source Links for remote starts"

Let's look at some Options

The focus of this book is around SMALL power planning – So I am going to focus on a couple of solutions that are on the smaller scale for power through a brownout. The requirements are:

- Portable Solution
- Lower Cost side
- Automatic Operation (Must keep our refrigerator and freezer running when not home)

We are going to first rule out some of our options we have spoken about and explain the reasons.

Whole Home Standby Generators

These tend to reside on the higher priced side of our options. You can get into a smaller system for a bit over $5000, but it is also a permanent solution and cannot be taken with you when you move. For renters this is not a good solution as it requires modifications to the home and the electrical system of the home we do not own. And for an apartment dweller or condominium and townhome owner is likely not a good solution due to permanence and fumes – no indoor operation.

Battery Based – Powerwall type of configuration

With installation most power walls will jump in price into the $10k to $15k range with only a couple of exceptions. Plus, we also are dealing with the issue of permanence – These are mounted and cannot be taken with you if you need to relocate. Installation also requires a professional and may require manufacturer specific transfer switches and interfacing equipment. One issue however we do get around that we have with any type of fuel generator – There are no fumes with a battery-based solution and these are all safe to install and operate indoors.

This leaves us with our good old trusty portable fuel and solar generators. The Fuel generator will ALWAYS be the price winner – with 9000-watt (Peak) Inverter generators available under $2,000 that will put out a constant 7000 watts of usable power EVERY HOUR we operate. That's a bunch of power – likely all we need for short term use.

There are the hurdles with a fuel generator though including:

- Auto start is not available on portable generators. You CAN however install one if your generator has a push-start option. This is a DIY modification, or you will need a generator expert to help you with it. See out list at the end of this chapter.
- Fumes – These still have fumes. If you use one it should be located outside or in a location where fumes cannot get into the home such as a shed or outdoor enclosure separated from the house.
- Fuel – Here the issue is dependency on fuel. You may have enough fuel to get through a outage of a few hours, but if you are away from home when the rolling blackouts occur, you won't be there to refuel the generator. (You could have a neighbor or family relative always check for you though – but this would still need someone to intervene).

That brings us to battery-based/solar generators. We have already ruled out Power-wall type of systems due to cost and permanence. But we have three larger portable systems which could fit our needs. Let's look at these now.

Goal Zero 6000x

This is a strong contender with 6000-watts of stored energy capacity and able to put out 2,000 watts of AC power – Plenty for refrigeration and freezer needs, medical equipment, and some lighting and appliances. At $4995 the cost is much more than a gas generator, but as it can charge while being used and it can easily power our devices while charging when the grid power it is running. It can also continue to provide power when the grid goes down. Provided the power is restored before we use out 6000-watts of power, the system will automatically recharge itself for the next rolling blackout cycle.

Pros:

- Portable design – Allows the unit to be moved if necessary. At a weight of about 100-lbs it is bulky, but it is on wheels.
- Pricewise, at $4995 with 6-kW of power capacity, the Goal Zero has more power per dollar than any other system out of the box. This is good but lack of expansion could be limiting for large-scale use.

Cons:

- Warranty of two years is the same as Point Zero's Titan, but less than that of any of our power walls.
- An expected battery life of 500-charge cycles is not that great in the world of solar. Now after the 500 cycles the battery will still store around 80% of its original capacity, so it is not a worn-out system. But with 500 cycles and 2-years this gives us a benchmark for comparisons and is less than all other systems we have looked at.
- If you need more than 6000-watts, there is no battery expansion available from Goal Zero.

Point Zero Titan Generator

The Point Zero Titan generator is a popular solar generator for off grid use and heavy applications. Like the Goal Zero it can put out a lot of power - 2500-watts of energy. Add two additional batteries and two 25-amp chargers for the batteries and you are priced at $4940 – in the same area as the Goal Zero. The Titan is an impressive system and deserves a good look at.

Pros:

- Expansion capability – you can stack up to six 2000-watt batteries based on their recommendation for up to 12,000 watts of stored energy.
- The Titan supports up to 2000-watts of solar panels allowing a rapid recharging of the system.
- Of our small portable systems we reviewed I would rank it the best as for expandability and capabilities, though for a while-home solution second to the Sol Donum.

Cons:

- Titan has a 24-month warranty (Which is not bad – but less than any of the 10-years of any of the Power Wall systems we reviewed.)
- The Battery life cycle is much better than Goal Zero's – with a LiFeP04 battery expected to last ten years. Charge cycles will be dependent upon how the Titan generator is used – High and fast power draws and high-power recharge cycles will short then life of the battery.
- One of the bigger drawbacks however – if you intend to use this as part of a permanent power solution for rolling blackouts – is a problem that the Goal Zero 6000x shares. Your appliances will be plugged into the Solar Generator ALWAYS drawing power from the inverter of the generator and the batteries while the chargers will continue to charge. This keeps the generator in a constant state of Charge/Discharge/Charge/Discharge which does place wear on the unit.

Sol Donum Vulcan

The last of the large battery-based solutions we must look at is from Sol Donum – A new company out of Virginia. They have hit the ball out of the park so to say with specifications that meet the need I am looking for. Their Charger/Inverter unit (1.4 kWh) coupled with their 3.6-kWh battery provides 5-kWh of power total with the ability add an almost unlimited number energy expansion units depending upon your specific configuration with additional Energy stores ("Energy Store" is the name they give to their battery expansion systems). Priced at $4890 their solution is in the same pricing level as both our prior portable solutions (Based on battery capacity) with features that neither of the other units have available such as pass-through power from the grid under non-blackout situations, and a new 240-volt model supporting higher voltage needs. This price is based on their newest models which omit the solar panel capability but making them perfect solutions where rolling blackouts are a problem. (Solar capable models are also available)

Pros:

- 2-kWh of AC Power out with a single Charger/Inverter unit, up to 4-kWh with two units in parallel. (More available with a custom configuration from Sol Donum)
- 10-year warranty – The only one of our portable solutions with such a warranty.
- **Pass-through Power from the Grid** (Huge advantage) – Instead of running off the batteries, you draw power from the grid until there is a power loss. Only then does the Vulcan automatically switch over to Battery/Inverter power and the switchover speed can be customized allowing immediate cutover, or delay. This reduces the charge/discharge wear on the batteries.
- LiFeP04 Battery chemistry – Currently the safest Lithium format and among the most popular.
- Each Charger/Inverter and each Energy Store comes with its own AC and optional MPPT Solar charger – So if you add additional energy stores you can add additional panels connected to the Energy Store without overloading the chargers or lengthening your recharge time.
- Designed, Engineered and Manufactured in the USA
- Internal Components can be replaced or upgraded if needed. Components such as the MPPT charger and the Inverter/Charger are off-the-shelf manufactured systems. This compared to other brands with specifically manufactured MPPT and Inverters to meet tight size specs.
- RFI & EMP Shielding available.
- Each unit is about 50-lbs allowing them to be movable. Bulky still, but about the size and weight of a 1500–2000-watt inverter generator.
- Portability and non-permanence add to viability as a solution for renters, apartment dwellers, or anyone where a fuel-based generator cannot be used due to fumes or flammable fuel.

- Can be used in an RV as a solution – best able to fit in a 5th wheel or larger motor home where often larger space areas are available.

Cons are:

- With new pricing for their Non-Solar model backup system – Ideal for rolling blackouts and short-term outages – and keeping in mind this is a mid-range power system optimized for higher power capacities, there are few cons to the Vulcan system. If you are looking for a smaller portable power system however, this is not your unit with the price for the solar model starting at $3495 for solar, $3095 for non-solar, there are much less expensive solutions available that are lighter and more portable. The strength of the Vulcan is in expansion and portability as a portable solution that can be expanded with 2.6-kW and 3.6-kW battery expansions that are competitively priced for their storage capacity.

Let's design some solutions

We're going to look at some solutions using Sol Donum's Vulcan solution as it is the most flexible of the manufactured solutions we have looked at. I am using their system as a model for three reasons.

1. Portability – Their system remains portable even if purchased and installed as a large-scale solution. Each Inverter/Charger unit and battery store is packaged as an individual unit weighing about 50 lbs. It is NOT permanently installed to a wall, and even can be installed into an RV as an RV solution. In an emergency, given about ten minutes, I could disconnect and move even a large 30-KW installed system (or larger) for moving if I had to.

2. Can operate either independently in a home or apartment powering from a wall outlet for charging, then passing through grid power to appliances when power is normal and will act as a UPS during a power outage while still supporting charging from an external source including the grid, a generator, or an array of solar panels.

3. It can also operate integrated into the home power panel using automatic transfer switches to supply a limited amount of power to multiple outlets through the house. (Limited to 2-kWh or 4-kWh depending on your configuration, though Sol Donum will build a larger custom solution for you if needed)

Full disclosure – Though I have no financial relationship with Sol Donum, I was fortunate enough to meet one of the Sol Donum team members during a training and speaking event where I was presenting in North Carolina. From that they have worked with me on questions and designs related to emergency power planning and have been a resource during the development of my updated training to include more in-depth rolling blackout protection planning.

I still have a strong belief that a hybrid solution is ultimately the best – A Fuel backup for peak needs and a battery-based backup to continue power on a more limited basis.

Using the system WITHOUT connecting to the Home Power infrastructure (Breaker Panel)

In this configuration we are going to set up the battery-generator in the home without connecting it through the home electrical panel. This method can work with any solar generator, but with the Vulcan we are going to assume a larger battery capacity with a 5-kWh battery capacity, plus also the fact that the Vulcan can pass through grid power while the power is on. This power-passing ability allows you to keep the batteries in a standby state without going through a discharge/charge cycle, and it almost instantly switches to battery when the power goes out. The auto cutover is part of what makes the Vulcan an ideal solution for rolling blackouts. After restoration of power the Vulcan will simply flip back to power-pass-thru operation and recharge from the wall charger at a rate of about 360 watts to be ready for the next blackout. In this configuration we are assuming we have regular daily power, so we recharge from the grid – no solar panels are necessary.

Below is a diagram provided by Sol Donum on using their system as an in-home power source without tying it directly into the home's power. This remains a simple DIY solution allowing you to place the Sol Donum device nearby the appliances you wish to power.

This configuration is simple – Just install your Vulcan Inverter/Charger unit into an AC wall outlet, and then plug your additional Battery Store into another wall outlet. This allows for charging of the batteries while grid power is active.

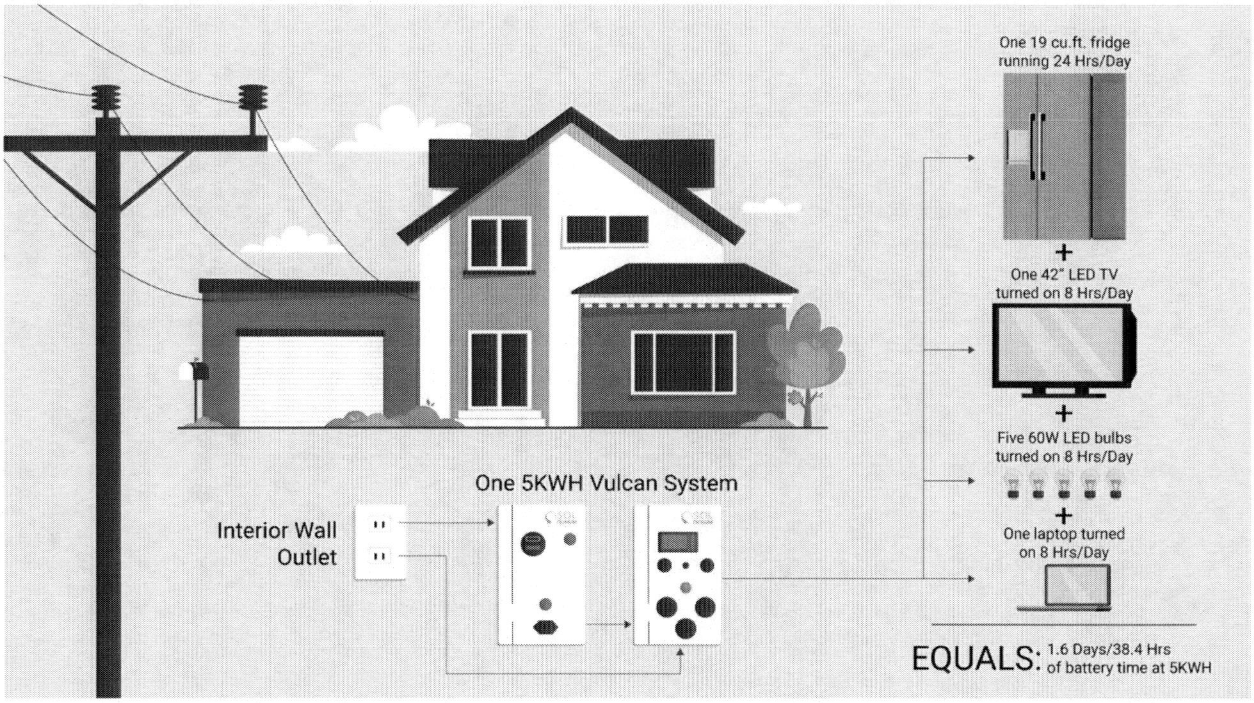

Next plug in any appliances that you wish operating during a power outage into the Inverter/Charger unit, and if necessary, install a power strip for additional outlets if you don't overload the 15-Amp limit on each outlet. This may include your refrigerator, lighting, or other appliances. Remember that while the grid is up and running all your devices are running off grid power passed through the Inverter/Charger unit.

That's all there is to it. Again – We can do this simple configuration with our Goal Zero and Titan generators, but with those we are running our appliances off our battery instead of with pass-through power. That's more effort on your batteries counting towards the charge/discharge cycles. If you own a smaller solution, such as an Apex of or one of the other solar generators we looked at in chapter three then these will also work but with limitations based on the battery capacity and inverter size limit. My Inergy Apex which has 1069-watts of storage capacity will last powering my refrigerator about eight hours. I can add additional 1-kW Batteries at a price of about $950 each. For "Automatic" operation however I would have to always leave those devices needing to be powered keeping my unit in that constant charge/discharge cycle.

Integrating into the Home Power

This configuration adds more complexity to our solution with the addition of a generator transfer switch. Though the installation of the Vulcan system remains as a DIY process, the connection to the home's breaker box and transfer switch should only be done by a licensed electrician. This additional transfer switch in this configuration DOES NOT HAVE TO BE AUTOMATIC. What we are doing here is leaving those home circuits that need to remain up during the blackout ALWAYS switched over.

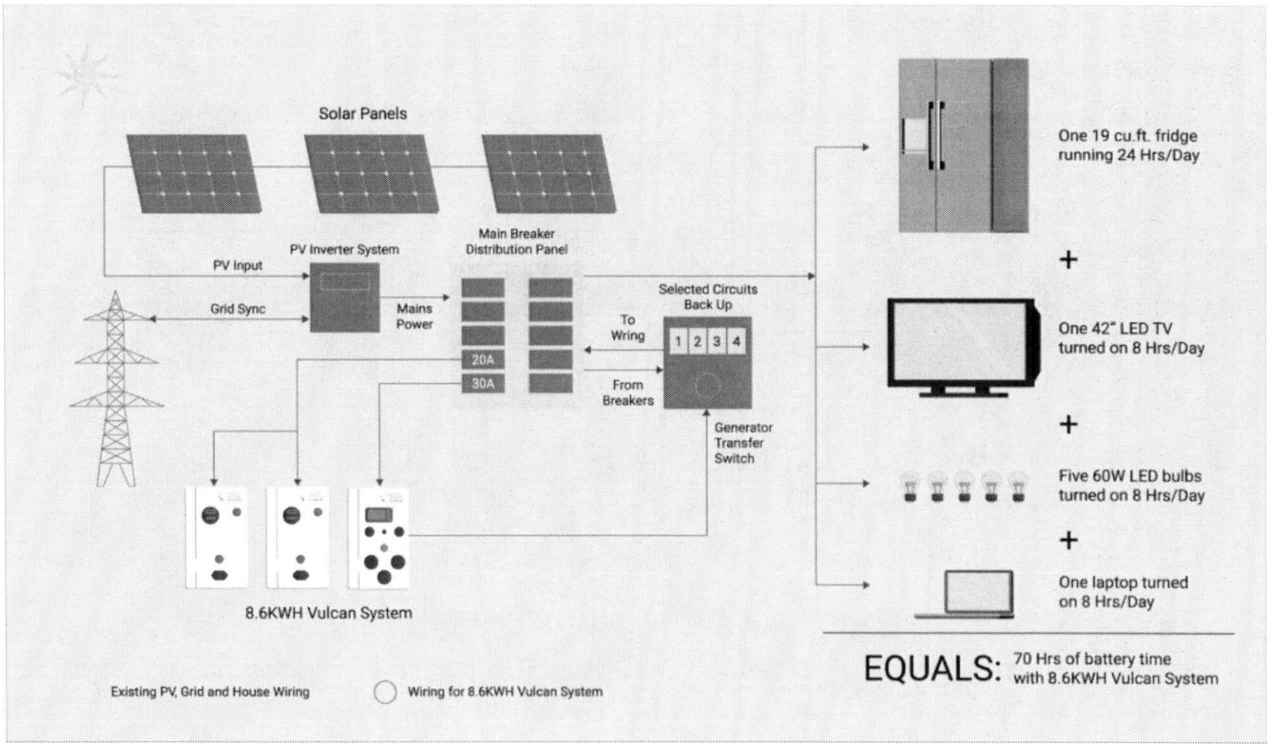

Normally when used with a backup gas generator, there would be no power to those circuits with the transfer switched in the fail-over mode since we would not start our generator until there was power loss. But – In this configuration remember that the Vulcan is going to pass through power then the grid is up, and therefore since it is the source of power for the transfer switch power will be running into the switch to power our circuits. When power fails, the Vulcan will automatically switch to the batteries and continue to provide power to the transfer switch, and thus to our devices. Anything running through the transfer switch will remain operational.

Our Vulcan has a 2000-watt power output capability, so we need to know that the combination of the power used by our outlets cannot exceed 2000-watts. Up to 4000-watts can be made available with a second Vulcan Inverter-Charger being added in parallel to the first.

In our diagram on the previous page (Provided by Sol Donum), the Vulcan system is charging from the wall, or will charge from optional solar panels. As in our prior example however, solar panels are not required to provide power during our rolling blackout window. The Inverter Charger/Battery store combination will recharge from AC power when the grid power is restored in about 11 hours (1.4-kWH battery + 3.6-kWH battery store). If your rolling blackout lasts between 4 and 8 hours, then this will be plenty of time to recharge before the next blackout cycle.

Integrating with Solar Panels

As discussed in both prior examples, solar panels are not necessary. Recharging from the AC Power when the grid is back up is simple and works well for anyone who cannot mount solar panels for energy – Apartment dwellers, home renters, or those living in condominiums, townhomes, or anywhere that mounted panels are not practical. However, if you want to be ready for a longer duration power outage and you are living in a home where you can mount panels then the Vulcan Solar capable model, or any of the solutions we have discussed in this chapter will easily support panel arrays. For the Vulcan the recommendation is to install an array of connected panels to each battery store or inverter charger allowing each to charge independently of the others. In the diagram below we see two 310-watt panels connected in series running to the Energy store, and two additional panels also in series to the Inverter/Charger.

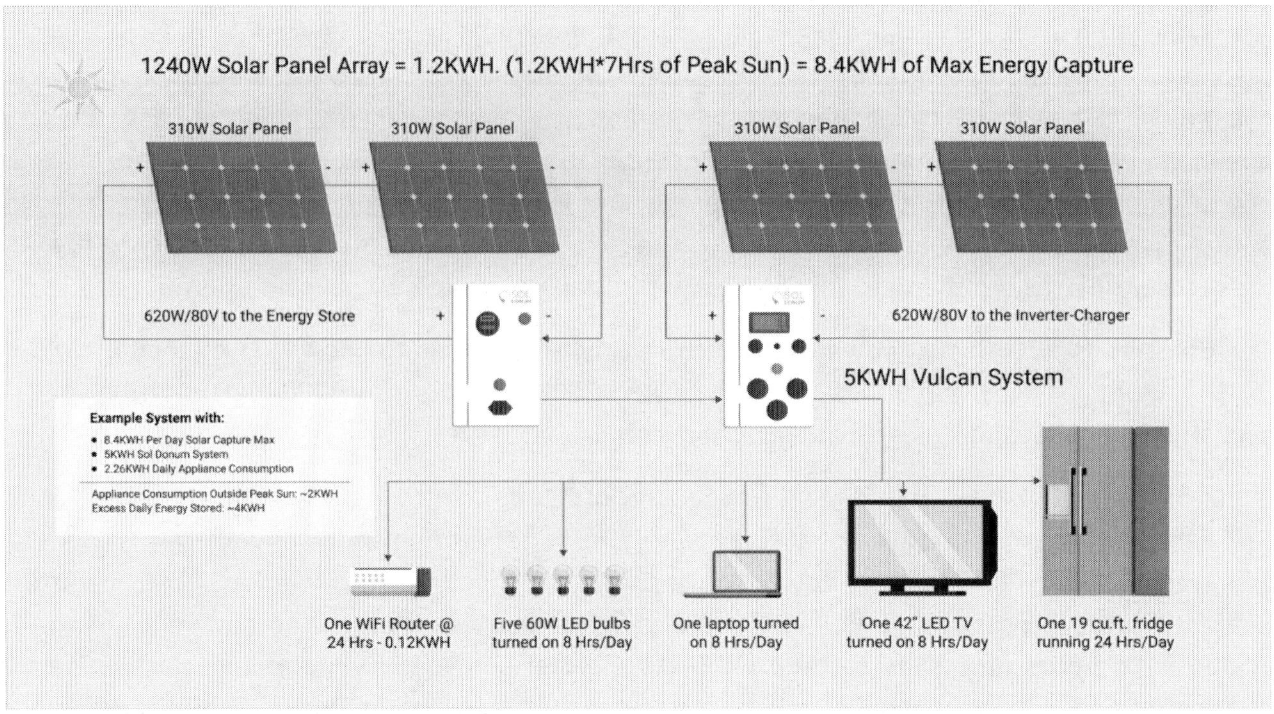

If during a power outage the grid stays down for longer than a few hours, such as following a hurricane, earthquake, or any other larger-scale emergency, then the ability to recharge and keep some power available remains. For myself, I still prefer to keep a high-capacity fuel generator available, but if fuel becomes limited this configuration will keep power to some lights and my refrigerator. This solution also serves to provide automatic power whereas my portable gas generator requires my intervention to start it up.

240-Volt Power Solutions – Well Pumps and other 240-volt Needs

A popular question that was asked during one of my seminars in a rural area was – what about well pumps?

If you have a need to power a well pump, most of which are 240-volt systems, or other 240-volt appliances, then your options are to purchase a larger fuel-based generator capable of putting out 240-volts which can be started and operated on an "As-needed" basis or use one of our battery-based solutions we have discussed. Several of the "Powerwall" class systems will handle 240-volt power out of the box. As of September 2021, Sol Donum will also have a 240-volt model of their Vulcan system available. As with all newer models they have both a Solar and a Non-solar model of their Inverter/Charger system available. If your installation need would be in a remote properly location such as a well house, barn, workshop or anywhere isolated then the Solar model at $3595 may be the best option with 1.4-kW of batter capacity able to recharge from one or more solar panels and run completely automated without the need to intervene and start up a generator.

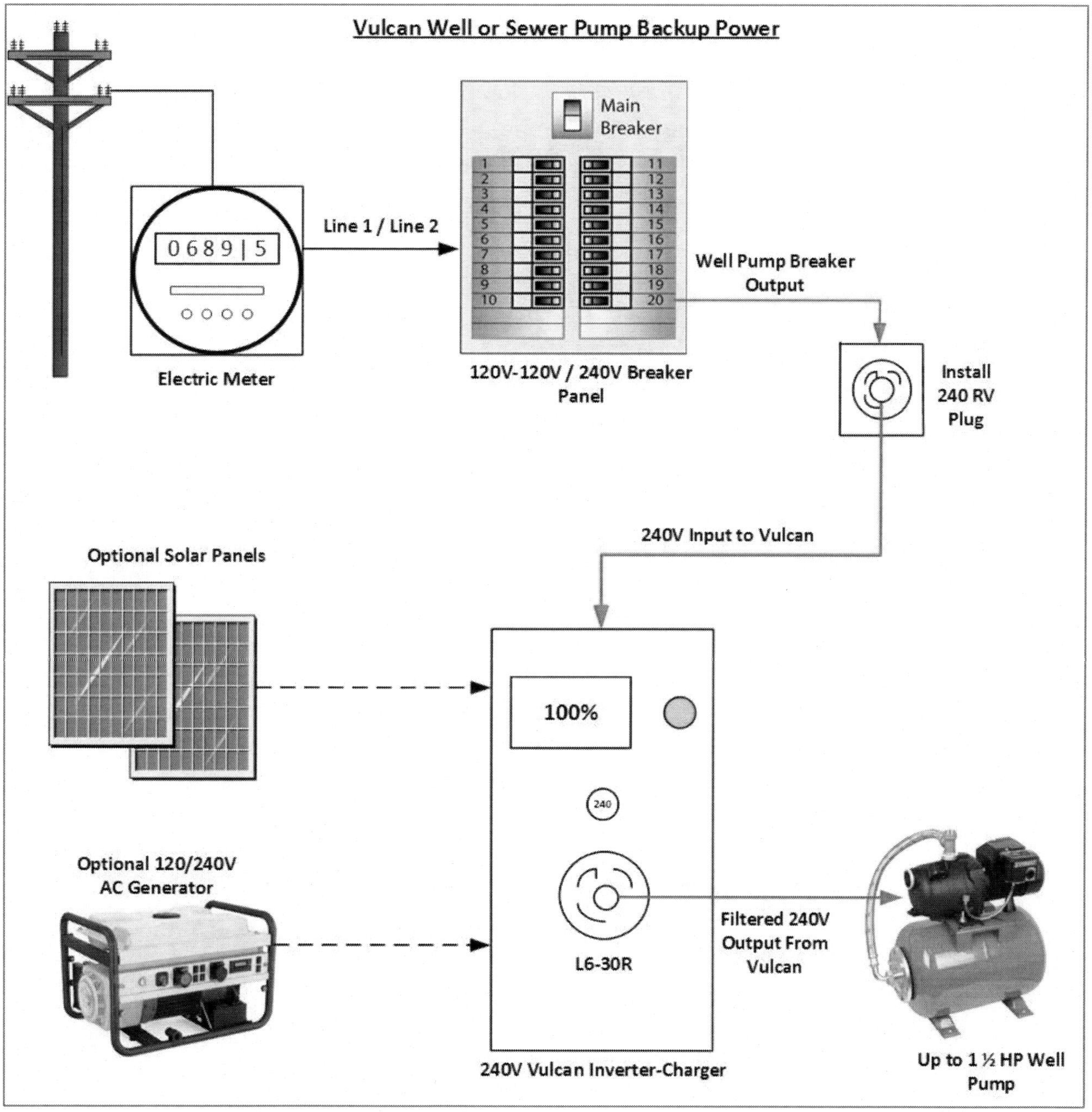

In the diagram above we see a Vulcan Inverter-Charger set up to provide power to a well pump. This configuration allows power to pass through during normal operation with 240-volts to the well pump. When the power goes out, power will automatically be provided by the 1.6-kWh battery in the Inverter-Charger system. If power returns such as during a rolling blackout then the Inverter-Charger will recharge from the grid. If power remains out for an extended period, then Solar Panels or an AC Generator can be used to recharge. If a 240-volt fuel generator is used, then while the generator is operating then that power will pass through to the well pump allowing the batteries to recharge until later needed.

Looking at a DIY solution

Beyond the pre-packaged solutions we have looked at in this chapter, you can also build a solution if you have the DIY skills. I have done this on a small scale for my camper and even a small 500-watt solution I have built for demonstration purposes that would easily keep my small appliances up and running if I needed to. We are not going to dig into detail on this as we have already touched on this in chapter 5, however there are some important points for a solution that will provide power when your regular AC power goes out.

Some differences to consider from a solar generator for an always "On" power backup include:

- Mounting – Mount on a board such as a section of plywood
- Look into an "Inverter/Charger" that provides PURE AC power out, and auto-switches to batteries, or passes through shore AC power as available.
  https://invertersrus.com/product-category/power-inverters/power-inverter-chargers/
  https://www.xantrex.com/power-products/inverter-chargers/overview.aspx
- Batteries will be a challenge based on your budget and what you can support. We recommend Lithium. 100-ah Lithium Batteries or larger will range widely in price from $500 (CHINS on Amazon) to Battleborn (Top quality about $950) to build-it-yourself batteries.
- Solar panels are only necessary if you need to charge from the sun.
- Building a Battery Solution – You can save a good amount by buying Lithium cells and building your own battery. This is not something I recommend for the novice not comfortable with electricity. Lithium batteries contain a large amount of power and shorting or mishandling these batteries can be a severe fire hazard and very dangerous.

SUMMARY

If rolling brownouts are a significant problem for you or in your area, then one of these solutions or a similar battery powered solution may be important to you. As mentioned in this chapter, battery-based solutions will be able to provide you a "Hot" solution able to automatically kick in when the power fails for a period of a few hours or even days with solar panels installed. These solutions will be much more expensive than a generator solution however will provide the ability to easily have backup power automatically covering you when you are away and protecting food, refrigeration, and any sensitive devices that would have problems during power loss.

Stories about Rolling Blackouts

- https://www.washingtonpost.com/weather/2021/01/19/high-winds-california-la-wildfires/
- https://www.mercurynews.com/2020/09/05/pge-warns-of-possible-power-shutoff-next-week-due-to-extreme-weather-high-wind/

Source Links for remote starts

- https://generatorupgradesupply.com/
- https://dcautogen.com/product/v2-autogen-12v-controller/
- https://www.pinellaspowerproducts.com/product/predator-3500-4-function-wireless-remote-control-kit/
- https://www.instructables.com/Autostart-House-Generator/ (Really DIY project)
- https://atkinsonelectronics.com/gscm-mini-60hz-usa-canada/

Reviews of Inverter/Chargers some of which with automatic Transfer

- https://www.galvinpower.org/best-automatic-transfer-switches/
- https://www.xantrex.com/power-products/inverter-chargers/overview.aspx
- https://www.victronenergy.com/inverters-chargers

9: What is YOUR Emergency Power Goal?

When you begin considering options for filling a power need, you will have to clarify what your goals are. There are many options that can fill power needs, from small to large solar solutions, and small to large generator options. Understanding your goals and needs will help you to identify the best solution for your specific need.

What is your Budget, and how much can you afford?
Probably one of the most important questions – What is your Budget and what cost can you afford? Overall a fuel generator will be the least cost. You can pick up a 700/900-watt Generator for under $200. This is a small, loud little generator that will provide some power. Then pick up some fuel cans and oil (This model requires mixing) and you're in business. For under $500 you can pick up either a small Inverter Generator under the 2000-watt range, or an open-frame non-inverter in a larger size.

As for Solar – You can get into a small solar generator, or build one, for $300-$500. Remember solar will be beneficial in a shelter or for using indoors. Add about $100 for a solar panel and you have a small renewable resource. If you cannot afford a larger solar system like this, then consider a small camping solar panel – you can use this to keep small devices charged via a USB port and keep your Cell phones and small USB batteries charged up.

Overall Solar will be the more expensive option of the two, however will not require gasoline which could be in limited supply. At the very least consider a small USB solar panel. A small panel like this can keep cell phones, radios, flashlights, and even handheld ham radios charged.

A whole-home power solution or a portable solution you can take with you?
Are you looking for a Whole-Home or Portable solution? A whole-home fuel-generator will set you back at least $4,000 to $5,000 installed and cannot be taken with you (not easily). But this is good if you intend to stay put and not move around, or if you need to power your home during blackouts that occur on a somewhat frequent basis such as rolling blackouts or remote areas where power lines may fail for extended periods due to weather or other events.

If your use is extremely occasional such as mine, or you need your power source to be portable such as I do then a smaller generator solution may be the better choice. Also, your lower power consumption and lower need may well justify a smaller and portable generator solution. For me – our neighborhood has a blackout that lasts over an hour only once every 3-4 years, and in ten years I have not had a blackout that has lasted over 4 hours. A whole-home generator would be extreme overkill in cost for me.

How long do you need backup power for? A few hours, days, or weeks?
The duration of your power needs will make a big difference in what solutions you chose. A smaller portable generator will work best for short-term periods. Remember a small generator should go through a "Cooling Off" window every eight to twelve hours, depending on model, and needs to be maintained with oil changes, spark plug changes, and filter cleaning periodically. You will also need to store fuel for multiple days. A standby generator running from Natural Gas will, most of the time, have natural gas available during a power outage (Some exceptions).

What type of fuel will you have available?
During a power event – A hurricane, a winter storm, flooding, or some other types of event, fuel may be difficult to come by. Gasoline is the most vulnerable of the fuels and depending on the type of event natural gas will most likely be the most available (For standby generators) but during the winter storm in Texas in 2021, supply of natural gas was short due to frozen pipelines and frozen distribution system. A dual fuel generator using gasoline and propane will offer flexibility. Also – during a hurricane from experience, gasoline runs out before diesel or propane.

Can you store enough fuel for the time you need the generator?
Fuel storage can be an issue both for safety and legal reasons in many areas. Of course, we don't want large volumes of flammable fuel for safety reasons, but also many communities have restrictions on the amounts of fuel you can store in your home. HOAs may also restrict homeowners from having large propane storage tanks in the community. Myself, though I don't keep a large amount of fuel at the house, but I do keep about ten 10-gallon gas cans empty and available. In Florida our most likely risk comes from Hurricanes which can be better planned for days in advance.

Will getting additional fuel be a problem?
Will you be able to get fuel during an emergency? In Florida we have had problems with gas stations running out before a hurricane leaving motorists unable to evacuate and travel if they are mid-way through an evacuation. Diesel tends to be available longer than gasoline, and then propane. For my gasoline portable generator, I try to use ethanol-free gasoline which is better for the small engine – this however will likely to be difficult to get during an emergency. During the 2021 Texas freeze propane became a problem as residents were turning to propane as a heat fuel source.

In addition to just being able to get the fuel, during many events power grids going down also kill the power needed to pump gasoline and diesel from service stations. Propane remains available as it is a pressurized fuel and power is not necessary to get the fuel from storage tanks where it is available for refill.

Do you need quiet or silent operation?
If you need silent operation, then you will need to look at either an Inverter generator or a solar generator. An inverter generator will still produce noise but at a much quieter level and will not be heard nearly as far as a contractor-style/open-frame generator. But for absolute silent

operation, a solar or battery-based solution is your best option. Other than a small cooling fan that may kick on in some models, solar generators create either a minor hum or no sound at all.

Do you need something you can use indoors?
If you need to have a power source for indoor use – such as for powering medical equipment or for use in an emergency shelter, then a solar generator is the only way to go. Any fuel-based generator will produce deadly fumes that build up when not ventilated. Never run a fuel generator indoors or even inside a garage. Always plan for proper ventilation. Fuel generators should operate several feet away from your house to ensure fumes do not rise above and into the roof overhangs.

Do you have sensitive devices to operate that can be affected by "Dirty" AC?
If you do have sensitive devices, then you need to look at either an Inverter style generator, or a solar generator with a pure-sinewave inverter built into it. The quality of the AC sinewave is measured by the "Total Harmonic Distortion" (THD) in the sinewave. A THD value that is 5% or less is considered safe where a THD greater than 5% is not. An inverter style generator will cost more than a non-inverter-based generator.

Do you need 240-volt power?
An often-overlooked question – Do you need to power a well pump, clothes dryer, or any device from 240-volts? Only the larger of the fuel-based generators have the capability to put out 240-volts of power. For a Solar or battery-based solution, only a few will have this ability. In most cases getting 240-volts to run a well will best be available from a larger fuel-based generator unless you go large with a larger battery solution such as the Sol Donum Vulcan, Tesla wall or Enphase power solution.

10: Balancing Systems for a Best Solution

For myself, I determined that my best solution would be a combination of having both a fuel generator and a small solar generator. You may also. I decided that a large solar generator would be just far too much in cost for the capabilities I needed. I had a need to get through a night with minimal power. The family can come together and slumber in the living room if we needed to. In my area power is reliable. We have no rolling blackouts, so the primary uses of a solar generator would be for emergency use, or camping either in my tent or my RV. When I tent camp, a single charge on the solar generator will usually last me 2-3 nights. In the RV I will mostly drain it overnight, but it recharges quickly with 500-watts of solar on the roof. (I carry 2 to 4 additional panels I can set up if I think my power need will be more).

If you have a need for running any type of medical equipment, a solar generator becomes much more valuable. You can run it indoors to operate medical monitors or CPAP equipment. No fumes and silent. Just be sure to size appropriately for your equipment. Most likely at least a 500-watt unit would work well for powering many CPAP or similar items.

If I am hunkering down after a storm, I do not want to attract attention from a loud running generator, and I will not sleep with a loud generator left outside and running. If camping, I opt to secure and lock down my fuel generator at night. I boondock – dry camp – on public lands often and the general rule is to shut off generators at night. So, when I do, the solar generator offers enough power to get through the night quietly. If the need to stay overnight in a shelter arises, then I can carry power with me for radio or gadget operation, running USB/LED lights, and keeping the phones charged.

I sized my solar generator to operate most needs inf my 30' camper with 30-amp electrical service in the camper. I can run a Microwave, lighting, radio, and television overnight with plenty of charge. If my camper were larger – say with a 50-amp service, and I traveled with a companion (Mostly solo here) then I would consider either more battery power or a unit capable of putting out more power – maybe closer to the 2000-watt size range.

But during an extended power outage – my solar generator will simply NOT be sufficient for my needs. It will not charge fast enough while under load for use more than a single night, and it will not provide enough power during the day if I need to run an Air Conditioner or heater. In fact – it will not provide sufficient power at night if I need cooling or heating either. If I needed a solar solution to run AC it would cost several times more for the battery storage – multiple thousands of dollars in batteries.

For day operation I use a 3500-watt Inverter generator. It is quieter and will idle down burning less power if I am not consuming a heavy load which saves fuel. 3000 running-watts is enough to power critical items in my home – refrigeration, a small AC room unit if needed, radio and television, ham radio, and some lights. Noise is not as much an issue during the day, as security is not also – not like it is at night. The inverter generators also produces a clean AC power source safe for sensitive devices. For many, 3000-watts may not be enough power. Portable inverter generators are available up to around 10,000-watts. If your devices do not need a pure sinewave you can easily find a much larger generator in the 5-kW to 10-kW range for under one to two thousand dollars from many good manufacturers.

The fuel generator frees my solar generator to simply recharge from panels during the day, or on an overcast day I can recharge the solar generator from the gas generator for the next evening. Of course, if security is not a concern, I can run the gas generator overnight – however as a gas generator does need a "Cool Down" window every few hours and the solar generator fills in the gap well. It also steps in when I need to do an oil change.

Apartment Dwellers

If you live in an apartment, I'd first consider a solar or battery generator as a backup power source – In the 2-kW range such as the Rich Solar unit. If you want a fuel generator for more power, remember the generator MUST be run outdoors away from the building several feet. (Not on the patio). This may work for someone on the 1st or even 2nd floor, but not in a tall building. A 2kW solar generator that can be recharged via AC and would provide power for your fridge and lights.

What would I purchase TODAY?

The generators I have for my own personal home use work well for me. I have a 3000/3500-watt Inverter based generator that will provide minimal power for my home during a sustained outage – refrigeration, some cooling, some heat, lighting, and other uses. For my RV Camper it is a 100% solution though if I had a larger camper with dual Air-Conditioners or requiring 50-amp service, it would fall short. For a larger camper, you would likely need a 4000 to 4500-watt fuel inverter generator.

If purchasing a new fuel generator today, I would be looking for a 2000-watt range small inverter generator that puts out a sustained 1800 to 2000-watts. Many small models only put out 1600-watts which is insufficient for running a 15,000 btu camper Air Conditioner. In addition, I would purchase a power parallel kit allowing me to connect a second generator of the same manufacturer and model for 4,000-watts of power. This solution also offers redundancy if I have a problem with one generator, allows me to carry a smaller and lighter weight generator for tent camping trips or smaller needs, and allows me to shut down one generator to rest or perform maintenance on while being able to depend on power from the other generator.

For solar, I have multiple solutions, but a solar generator in the 1500-watt to 2000-watt range with expandable battery capacity and at least 500-watts of solar input I consider to be the minimal. This offers me plenty of power for a small dorm-style refrigerator (for medication), some lighting, radio/television, and ham radio operation. My Inergy Apex meets these needs, and there are several other good batter/solar based generators on the market that would also. I also have smaller solar generators I have built (See chapter 7) as well as a solar capability in my camper that provides me a small power capability without the solar generator or acts as the high wattage charging system when my solar generator is plugged into the camper.

Following is a list of common solutions with cost estimations. Note these are not specific to any manufacturer and are general. Some of the low-end Solar solutions are DIY based.

Solution	Est Cost	Notes
Gas Generator - 2-kW, Inverter Generator	$500 - $1100	2-kW provides the lowest general generator power size for use, and an Inverter provides a clean pure sinewave for best quality power.
Gas Generator - Non-Inverter	$400 - $2500 or more	This covers a wide range of sizes and brands - be watchful of cheap generators that could be problematic. Honda generators have a strong reputation and highest cost. There are many good lower priced models. Non-inverter provides a "Modified" sine wave- not as ideal for many electronics and devices. Also, generally louder, and often less fuel efficient.
Solar - Small (Renogy or Harbor-Freight Solution, DIY based)	> $300	For less than $300 you can set up a quick and simple solution with everything purchased from Amazon or Harbor Freight - Giving you about 300-400 watts or usable AC power and 100-watts of solar panels useful to keep phones and items charged, a radio or tv working, and even a ham radio operating.
Small Solar- DIY - 500 Watts, With panels	$500 - $700	Most are priced without panels so take that into account. Solar produces a fraction of the power of a Gas generator at a higher cost, but can be used indoors, fumeless, no fuel issues, no maintenance.
Medium Solar – 1Kw with Panels	$1500-$2000	"Medium" sized units last longer, provide more power, can charge faster. Note prices may jump up with more panels - this price estimate is based on one or two panels, but many generators will support up to five or more panels for fastest recharging of the battery.
Large Solar – 3-kW with Panels (Portable, Solar Generators)	$3800 - $8000	This range varies greatly - these sized "Portable" solar generator kits can provide power to much of your home but could also be portable for use elsewhere. Smallest example is a Titan Solar Generator – 2kW with 500-watts of panels. Largest is 6-kW of Battery, with 2kW of Solar Panels. Other manufacturers including Goal Zero have products within this range.
Full-Home Solar, Permanent Mounting	$5k - $30k+	For a full-home Solar solution we are casting here a wide net. There are so many questions that must be asked, it can be a huge difference between solutions. If you are looking for a simple solution to power an off-grid cabin with light use, then a solar solution may be very reasonable, as little as the $5k range or even less depending upon needs. But – If you are looking for a large-scale system to power the entire house for days or weeks, then a very careful analysis is needed to cover the needs you have.

11: Power Planning – Tables, Charts & Notes

We have talked in the previous chapters topics covering fuel based and battery-based power systems. There are a wide range of specifications that need to be considered when deciding on the solution that best fits your specific need. The very first thing you need to decide on is how much power do you need? The amount of power that you need will be dependent on the devices that you need to operate, and the devices you need to operate are dependent on how long you are planning to be off grid for.

In the tables appearing on the following pages, we have compiled wattage estimates from a series of different sources and combined into a list of over 300 items. The estimated wattages that are listed are just that – estimates. There are only two ways to get accurate wattage requirements for your devices – one is to look at the manual or specifications published by the manufacturer. The other is with the use of a Wattmeter which we will cover later in this chapter. The wattmeter is a measurement device that plugs into the wall and then has a 120-volt outlet on it that you plug your device into. With it being in-between the power source and your device, it can easily and reliably make the measurements and see exactly what is being consumed.

The Wattage Estimates tables on the next pages are estimates only. You will notice that for many of the items there are several variations. For Air Conditioners for instance, we have 18 variations based on type of AC unit and size. Refrigerators have a dozen variations based on type and size, and coffee makers come in nine separate flavors. Even when you find the device that most closely matches what you own – be aware that these estimates can be inaccurate, and you should validate your own device's power consumption needs from your owner's manual or through a wattmeter.

Wattage Estimates by Device

The following tables contain over 300 various items and estimated wattage use per hour for each of the devices. The wattage values given are only estimates – but can be used to help develop a load estimate for how much power you need to plan for daily. For each item multiply the Watts/Hour by the number of hours or fractional hours each day you will need. For some devices such as refrigerators we have already calculated out daily power needs. For other devices, such as microwave ovens, calculate the watts/hour by the fractional hours to determine the power need. (A 1500-watt Microwave used for 20 minutes/day will require 1500/3 watts, or 500-watts where the 3 divisor represents one-third of an hour, or 20 minutes)

AC/Cooling

Ref #	Category	Sub-Category	Item	Watts/Hr	Source	Notes
1	AC/Cooling	AC	1.5-ton AC Unit	2930	(5)	
2	AC/Cooling	AC	5.0 ton	9770	(5)	
3	AC/Cooling	AC	Air Conditioner - Central	10000	(7)	
4	AC/Cooling	AC	Air Conditioner - Room	1500	(7)	
5	AC/Cooling	AC	Air Conditioner – RV	2000	(7)	
6	AC/Cooling	AC	Air Conditioner (12,000 BTU)	1500	(6)	
7	AC/Cooling	AC	Air Conditioner (36,000 BTU)	4500	(6)	
8	AC/Cooling	AC	Air Cooler	80	(4)	
9	AC/Cooling	AC	Air Purifier	30	(4)	
10	AC/Cooling	AC	Central (3 ton–12 SEER)	3000	(5)	
11	AC/Cooling	AC	Window AC 10.000 BTU	1200	(8)	
12	AC/Cooling	AC	Window AC 12.000 BTU	1500	(8)	
13	AC/Cooling	AC	Window AC 15.000 BTU	1900	(8)	
14	AC/Cooling	AC	Window AC 5.000 BTU	600	(8)	
15	AC/Cooling	AC (RV)	RV Air Conditioner	900	(8)	
16	AC/Cooling	AC (RV)	RV Air Conditioner	1200	(8)	
17	AC/Cooling	AC (RV)	RV Air Conditioner	1600	(8)	
18	AC/Cooling	AC (RV)	RV Air Conditioner	1700	(8)	
19	AC/Cooling	Dehumidifier	Dehumidifier	280	(3)	
20	AC/Cooling	Fan	Box Fan	200	(1)	
21	AC/Cooling	Fan	Ceiling Fan	120	(3)	
22	AC/Cooling	Fan	Fan – Portable	150	(7)	
23	AC/Cooling	Fan	Fan (Attic)	400	(6)	
24	AC/Cooling	Fan	Pedestal Fan	60	(4)	
25	AC/Cooling	Fan	Table Fan	75	(9)	
26	AC/Cooling	Fan	Tower Fan	60	(4)	
27	AC/Cooling	Fan	Wall Fan	60	(4)	
28	AC/Cooling	Fan	Whole house fan	400	(5)	

Auto

29	Auto	Auto	Auto Engine Heater	600	(6)	
30	Auto	Auto	EV Car Charger	7000	(4)	
31	Auto	Battery Charger	Battery Charger (Car)	150	(6)	

Bathroom

32	Bathroom	Curling Iron	Curling Iron	150	(8)	
33	Bathroom	Hair Dryer	Hair Dryer	1500	(3)	
34	Bathroom	Hair Rollers	Heated Hair Rollers	400	(4)	
35	Bathroom	Heated Mirror	Heated Bathroom Mirror	70	(9)	
36	Bathroom	Razor	Electric Razor	20	(9)	
37	Bathroom	Shower	Power Shower	7500	(9)	
38	Bathroom	Towel Heater	Bathroom Towel Heater	150	(4)	

Computers

39	Computers	Charger	Tablet Charger	10	(9)	
40	Computers	Desktop	Desktop Computer	100	(9)	
41	Computers	Desktop	Desktop Computer (Gaming)	500	(1)	
42	Computers	Desktop	Desktop Computer (Standard)	200	(1)	
43	Computers	Desktop	Desktop computer on sleep/standby mode	6	(5)	
44	Computers	Desktop	iMac	240	(4)	
45	Computers	Laptop	Chromebook	45	(4)	
46	Computers	Laptop	Laptop	100	(1)	
47	Computers	Laptop	Tablet Computer	10	(4)	
48	Computers	Laptop	Thin client	6	(5)	
49	Computers	Monitor	Computer Monitor	30	(4)	
50	Computers	Monitor	LCD Monitor	100	(3)	
51	Computers	Printer	Laser Printer	800	(4)	

Electronics

52	Electronics	Cable Device	Apple TV	6	(4)	
53	Electronics	Charger	Cell Phone Charger	5	(9)	
54	Electronics	Charger	Phone Charger	7	(4)	
55	Electronics	Clock	Clock	3	(6)	
56	Electronics	Clock	Clock Radio	7	(3)	
57	Electronics	DVD	Bluray Player	15	(3)	
58	Electronics	DVD	DVD Player	26	(9)	
59	Electronics	Echo	Amazon Echo	3	(4)	
60	Electronics	Gaming	Game Console	200	(4)	
61	Electronics	Gaming	Gaming PC	700	(9)	
62	Electronics	Gaming	Nintendo Switch	18	(5)	
63	Electronics	Gaming	Nintendo Switch AC Adapter	40	(4)	
64	Electronics	Gaming	Play Station 4	90	(4)	
65	Electronics	Gaming	Video Game Console	150	(3)	
66	Electronics	Gaming	Xbox One	110	(4)	
67	Electronics	Modem	Modem	7	(3)	

68	Electronics	Phones	Cordless Phone	2	(9)	
69	Electronics	Phones	Home Phone	5	(4)	
70	Electronics	Phones	Smart Phone – Recharge	6	(3)	
71	Electronics	Phones	Telephone – Radio	150	(7)	
72	Electronics	Roku	Roku	3	(5)	
73	Electronics	Routers	Home Internet Router	15	(4)	
74	Electronics	Routers	Router/DSL/Cable Modem	6	(2)	
75	Electronics	Routers	WiFi Booster	2	(4)	
76	Electronics	Routers	WiFi Router	10	(4)	
77	Electronics	Sound Systems	Home Sound System	95	(4)	
78	Electronics	Sound Systems	Home Theatre	315	(9)	

Heating

79	Heating	Heaters	Electric heater (1500 W)	1500	(5)	
80	Heating	Heaters	Electric Heater Fan	2000	(9)	
81	Heating	Heaters	Fan – Furnace	1000	(7)	
82	Heating	Heaters	Furnace Fan 1/2 HP	800	(8)	
83	Heating	Heaters	Furnace Fan 1/3 HP	700	(8)	
84	Heating	Heaters	Furnace Fan Blower	800	(1)	
85	Heating	Heaters	Heat Lamp	250	(6)	
86	Heating	Heaters	Heat pump (>75 gal) water heater	155	(5)	
87	Heating	Heaters	Heat pump (50–75 gal) water heater	107	(5)	
88	Heating	Heaters	Heat pump dryer	540	(5)	
89	Heating	Heaters	Heat pump heat strips	10000	(5)	
90	Heating	Heaters	Heater – Electric	1500	(7)	
91	Heating	Heaters	Space Heater 2 kW	2000	(8)	

Household

92	Household	Air Cleaner	Electronic Air Cleaner/Filter	50	(2)	
93	Household	Aquarium	Aquarium	200	(8)	
94	Household	Aquarium	Aquarium Pump	50	(4)	
95	Household	Electric Blanket	Electric blanket: Double/Queen	94	(5)	
96	Household	Electric Blanket	Electric blanket: King	125	(5)	
97	Household	Electric Blanket	Electric blanket: Twin	63	(5)	
98	Household	Exhaust Fan/Blower	Exhaust Blower	150	(9)	
99	Household	Exhaust Fan/Blower	Extractor Fan	12	(4)	
100	Household	Furnace	Electric Furnace	20000	(9)	
101	Household	Garage Door	Garage Door	550	(9)	

102	Household	Garage Door	Garage Door Opener	350	(6)	
103	Household	Humidifier	Humidifier (Winter)	177	(6)	
104	Household	Humidifier	Humidifier 13 gal.	175	(8)	
105	Household	Sewing Machine	Sewing Machine	100	(9)	
106	Household	Vacuum	Dyson vacuum cleaner	230	(5)	
107	Household	Vacuum	Robot vacuum cleaner	7	(5)	
108	Household	Vacuum	Vacuum	1000	(3)	
109	Household	Vacuum	Vacuum Cleaner	1560	(6)	
110	Household	Vacuum	Vacuum Cleaner	1225	(7)	
111	Household	Water Dispenser	Hot Water Dispenser	1300	(4)	
112	Household	Water Dispenser	Hot Water Immersion Heater	3000	(9)	
113	Household	Water Heater	Electric Tankless Water Heater	8800	(4)	
114	Household	Water Heater	Electric Thermal Radiator	500	(4)	
115	Household	Water Heater	Electric water heater	694	(5)	
116	Household	Water Heater	Electric Water Heater 4 kW	4000	(8)	
117	Household	Water Heater	Tankless Water Heater – Electric	18000	(3)	
118	Household	Water Heater	Water Heater – Electric	4500	(3)	
119	Household	Water Heater	Water Heater (Quick Recovery)	4500	(6)	
120	Household	Water Heater	Waterbed Heater (Queen Size)	375	(6)	

Kitchen

121	Kitchen	Appliances	Cooker Hood	30	(4)	
122	Kitchen	Appliances	Electric Can Opener	170	(8)	
123	Kitchen	Appliances	Food Blender	400	(4)	
124	Kitchen	Appliances	Food Dehydrator	800	(9)	
125	Kitchen	Appliances	Food Processor	400	(8)	
126	Kitchen	Appliances	Hot Water Dispenser	1200	(9)	
127	Kitchen	Blender	Blender	500	(1)	
128	Kitchen	Broiler	Broiler, portable electric	1500	(5)	
129	Kitchen	Coffee Maker	Coffee Maker	1750	(7)	
130	Kitchen	Coffee Maker	Coffee maker/brew, warmer on	400	(5)	
131	Kitchen	Coffee Maker	Coffee Maker: Drip (Brew)	1500	(2)	
132	Kitchen	Coffee Maker	Coffee Maker: Drip (Warm)	70	(2)	
133	Kitchen	Coffee Maker	Coffee Maker: Percolator (Brew)	600	(2)	
134	Kitchen	Coffee Maker	Coffee Maker: Percolator (Warm)	80	(2)	

135	Kitchen	Coffee Maker	Coffee Roaster	600	(9)	
136	Kitchen	Coffee Maker	Espresso Coffee Machine	1500	(4)	
137	Kitchen	Coffee Maker	Percolator	1100	(4)	
138	Kitchen	Compactor	Compactors	400	(2)	
139	Kitchen	Cooking	Convection Oven	1500	(6)	
140	Kitchen	Cooking	Cooking Stove Top	1500	(9)	
141	Kitchen	Cooking	Electric Barbeque Grill	1560	(9)	
142	Kitchen	Cooking	Electric Boiler	14000	(4)	
143	Kitchen	Cooking	Electric Grills	1800	(2)	
144	Kitchen	Cooking	Electric Kettle	1200	(9)	
145	Kitchen	Cooking	Electric Pressure Cooker	1000	(4)	
146	Kitchen	Cooking	Electric Rice Cooker	1800	(9)	
147	Kitchen	Cooking	Electric Skillet	1500	(9)	
148	Kitchen	Cooking	Electric stove	2000	(4)	
149	Kitchen	Cooking	Fry Pan	1200	(6)	
150	Kitchen	Cooking	Fryer	1000	(4)	
151	Kitchen	Cooking	Induction Cooker	2900	(9)	
152	Kitchen	Cooking	Kettle – Electric	1200	(3)	
153	Kitchen	Cooking	Oven	2300	(5)	
154	Kitchen	Cooking	Oven: self-cleaning feature	6000	(5)	
155	Kitchen	Cooking	Oven: surface	1500	(5)	
156	Kitchen	Cooking	Pizza Maker	1440	(9)	
157	Kitchen	Cooking	Popcorn Maker	1500	(9)	
158	Kitchen	Cooking	Pressure Cooker	700	(4)	
159	Kitchen	Cooking	Rice Cooker	800	(4)	
160	Kitchen	Cooking	Slow Cooker	270	(7)	
161	Kitchen	Dishwasher	Dishwasher	1800	(7)	
162	Kitchen	Dishwasher	Dishwasher (Dry Cycle)	1200	(6)	
163	Kitchen	Dishwasher	Dishwasher (Wash Cycle)	200	(6)	
164	Kitchen	Dishwasher	Dishwasher: Energy-saver cycle	500	(5)	
165	Kitchen	Dishwasher	Dishwasher: normal cycle (not including hot water)	2000	(5)	
166	Kitchen	Disposal	Garbage Disposal	450	(3)	
167	Kitchen	Filter	Water Purifier	25	(9)	
168	Kitchen	Freezer	Freezer – Chest – 15 cu. ft.	45	(3)	
169	Kitchen	Freezer	Freezer – Upright – 15 cu. ft.	52	(3)	
170	Kitchen	Freezer	Freezer - Upright - 15 cu. ft. (2)	52	(1)	
171	Kitchen	Grills	Sandwich Grill	700	(9)	
172	Kitchen	Grills	Sandwich Maker	1000	(4)	
173	Kitchen	Grills	Waffle Iron	1200	(6)	
174	Kitchen	Heaters	Air Fryer	1500	(4)	
175	Kitchen	Ice	Ice Maker	100	(9)	

176	Kitchen	Microwave	Microwave - 1000 Watt	1000	(3)	
177	Kitchen	Microwave	Microwave- 1500 Watt	1500	(7)	
178	Kitchen	Refrigerator	ENERGY STAR Refrigerator (frost-free) 24 cu. ft.	75	(5)	
179	Kitchen	Refrigerator	ENERGY STAR Refrigerator (frost-free), 17 cu. ft.	49	(5)	
180	Kitchen	Refrigerator	ENERGY STAR Refrigerator (frost-free), 19 cu. ft.	64	(5)	
181	Kitchen	Refrigerator	ENERGY STAR Refrigerator (side by side) 21 cu. ft.	71	(5)	
182	Kitchen	Refrigerator	ENERGY STAR Refrigerator (side by side) 25 cu. ft.	83	(5)	
183	Kitchen	Refrigerator	ENERGY STAR Refrigerator, 14 cu. ft.	48	(5)	
184	Kitchen	Refrigerator	Refrigerator - 2.0 cu Dorm (AC) (2)	34	(1)	
185	Kitchen	Refrigerator	Refrigerator – 20 cu. ft. (AC)	59	(3)	
186	Kitchen	Refrigerator	Refrigerator - 20 cu. ft. (AC) (2)	59	(1)	
187	Kitchen	Refrigerator	Refrigerator (frost-free), 15 cu. ft.	208	(5)	
188	Kitchen	Refrigerator	Refrigerator / Freezer (Average)	150	(9)	
189	Kitchen	Refrigerator	Refrigerator -16 cu. ft. (AC)	50	(3)	
190	Kitchen	Refrigerator	Wine Chiller	200	(9)	
191	Kitchen	Refrigerator	Wine cooler (18 bottles)	83	(9)	
192	Kitchen	Toaster	Bread Toaster	1200	(9)	
193	Kitchen	Toaster	Toaster	1400	(6)	
194	Kitchen	Toaster	Toaster Oven – Oven	1500	(2)	
195	Kitchen	Toaster	Toaster Oven -Toasting	460	(2)	
196	Kitchen	Water Cooler	Water Cooler	100	(9)	

Laundry

197	Laundry	Dryer	Clothes dryer (light load vs. heavy load)	3250	(5)	
198	Laundry	Dryer	Dryer - Clothes, Electric	5400	(7)	
199	Laundry	Dryer	Dryer - Clothes, Gas	400	(7)	
200	Laundry	Iron	Clothes Iron	1100	(2)	
201	Laundry	Iron	Steam Iron	2500	(4)	
202	Laundry	Steamer	Garment Steamer	1600	(9)	
203	Laundry	Steamer	Hot wash, warm rinse	6300	(5)	
204	Laundry	Washer	ENERGY STAR-Qualified Washing Machine	1900	(5)	
205	Laundry	Washer	Non ENERGY STAR-Qualified Washing Machine	2280	(5)	
206	Laundry	Washer	Washer	512	(6)	
207	Laundry	Washer	Washing Machine	1200	(8)	

Lighting

208	Lighting	Bulbs	100W light bulb (Incandescent)	100	(9)	
209	Lighting	Bulbs	60W light bulb (Incandescent)	60	(4)	
210	Lighting	Bulbs	CFL Bulb – 100 Watt Equivalent	30	(3)	
211	Lighting	Bulbs	CFL Bulb – 40 Watt Equivalent	11	(3)	
212	Lighting	Bulbs	CFL Bulb – 60 Watt Equivalent	18	(3)	
213	Lighting	Bulbs	CFL Bulb – 75 Watt Equivalent	20	(3)	
214	Lighting	Bulbs	CFL Light Bulb	14	(9)	
215	Lighting	Christmas Lights	LED Christmas Lights	5	(4)	
216	Lighting	Fluorescent	Compact Fluorescent 20 Watt	22	(1)	
217	Lighting	Fluorescent	Compact Fluorescent 25 Watt	28	(1)	
218	Lighting	Fluorescent	Lighting (Compact Fluorescent)	18	(6)	
219	Lighting	Fluorescent	Lighting (Fluorescent)	40	(6)	
220	Lighting	Halogen	Halogen – 40 Watt	40	(3)	
221	Lighting	Halogen	Halogen (300 W)	300	(5)	
222	Lighting	Halogen	Halogen Work Light 1 kW	1000	(8)	
223	Lighting	Halogen	Halogen Work Light 300 W	300	(8)	
224	Lighting	Halogen	Halogen Work Light 500 W	500	(8)	
225	Lighting	LED Bulb	CFL/LED (11 W) equivalent to 40 W incandescent	10	(5)	
226	Lighting	LED Bulb	CFL/LED (15 W) equivalent to 60 W incandescent	15	(5)	
227	Lighting	LED Bulb	CFL/LED (20 W) equivalent to 75 W incandescent	20	(5)	
228	Lighting	LED Bulb	CFL/LED (27 W) equivalent to 100 W incandescent	27	(5)	
229	Lighting	LED Bulb	CFL/LED (38 W) equivalent to 150 W incandescent	38	(5)	
230	Lighting	LED Bulb	CFL/LED (8 W) equivalent to 25 W incandescent	8	(5)	

Medical

231	Medical	Medical	Nebulizer	1000	(5)	
232	Medical	Medical	Oxygen concentrator	460	(5)	
233	Medical	Medical	Sleep apnea machine (CPAP)	200	(5)	
234	Medical	Medical	Sterilizer	650	(4)	
235	Medical	Medical	Kidney Dialysis	1450	(10)	
236	Medical	Medical	Specialty Adjustable Bed	85	(10)	

Office

237	Office	Copiers	home copiers	800	(2)	
238	Office	Fax	Home fax/Multi-function device (inkjet)	18	(2)	
239	Office	Printer	Printers (Inkjet)	13	(2)	
240	Office	Printer	Printers (Laser)	250	(2)	
241	Office	Projector	Projector	220	(9)	
242	Office	Scanner	Scanner	10	(9)	
243	Office	Shredder	Paper Shredder	200	(9)	

Radios

244	Radios	CB Radio	CB Radio - Transmit @ 4 watts	6	(1)	
245	Radios	CD Player	CD Player	7	(2)	
246	Radios	CD Player	CD, Tape, Radio, Receiver System	250	(6)	
247	Radios	CD Player	CD/DVD Player	100	(8)	
248	Radios	Clock Radio	Radio – Clock	10	(7)	
249	Radios	Clock Radio	Radio, CD player	20	(5)	
250	Radios	DVD	Bluray/DVD Player (1)	15	(1)	
251	Radios	Ham Radio	Ham Radio - Transmit @ 50 watts	75	(1)	
252	Radios	Radio	AM/FM Radio	100	(8)	
253	Radios	Radio	Boom Box	8	(2)	
254	Radios	Radio	Radio	71	(6)	
255	Radios	Radio	Radiotelephone – Receive	5	(3)	
256	Radios	Radio	Radiotelephone – Transmit	75	(1)	
257	Radios	Stereo	Stereo	250	(8)	

Televisions

258	Televisions	Cable Box	Cable box	16	(5)	
259	Televisions	CRT TV	TV (CRT)	60	(2)	
260	Televisions	DLP TV	TV (DLP)	175	(2)	
261	Televisions	LCD TV	TV (LCD)	150	(2)	
262	Televisions	LED TV	22 Inch LED TV	17	(4)	
263	Televisions	LED TV	32 Inch LED TV	60	(4)	
264	Televisions	LED TV	42 Inch LED TV	60	(4)	
265	Televisions	LED TV	46 Inch LED TV	70	(4)	
266	Televisions	LED TV	49 Inch LED TV	85	(4)	
267	Televisions	LED TV	55 Inch LED TV	116	(4)	
268	Televisions	LED TV	65 Inch LED TV	130	(4)	
269	Televisions	LED TV	82 Inch LED TV	295	(4)	
270	Televisions	OLED TV	OLED/4K >40"	140	(5)	
271	Televisions	OLED TV	OLED/4K 20"–30"	23	(5)	

272	Televisions	OLED TV	OLED/4K 30"–40"	45	(5)	
273	Televisions	Plasma TV	Plasma TV >50"	480	(5)	
274	Televisions	Plasma TV	Plasma TV 40"–49"	400	(5)	
275	Televisions	Plasma TV	TV – Plasma (Small - 20" to 32")	200	(3)	
276	Televisions	Projection TV	TV (CRT - Projection)	225	(2)	
277	Televisions	Satellite	Satellite Receiver	250	(7)	
278	Televisions	Television	25" colour TV	150	(9)	
279	Televisions	VCR	VCRs	18	(2)	

Tools

280	Tools	Air Compressor	Air Compressor 1 HP	1600	(8)	
281	Tools	Air Compressor	Air Compressor 1/4 HP	900	(8)	
282	Tools	Blower	Corded Electric Handheld Leaf Blower	2500	(4)	
283	Tools	Charger	Cordless Drill Charger	70	(9)	
284	Tools	Drill	Drill – 1/2"	750	(3)	
285	Tools	Drill	Drill - 1/4"	250	(1)	
286	Tools	Drill	Drill – 1"	1000	(3)	
287	Tools	Drill	Electric Drill 1/2 inch 5.4 Amps	600	(8)	
288	Tools	Drill	Electric Drill 3/8 inch 4 Amps	450	(8)	
289	Tools	Drill	Hammer Drill 1.2 kW	1200	(8)	
290	Tools	Grinder	Wet Grinder	300	(9)	
291	Tools	Jointer	Planer/Jointer 6 inch	1800	(8)	
292	Tools	Mixer	Mixer Grinder	750	(9)	
293	Tools	Mixer	Mixer, Hand	100	(6)	
294	Tools	Mower	Electric Mower	1500	(4)	
295	Tools	Mower	Mower (Corded, 14 Amps)	1700	(8)	
296	Tools	Paint Sprayer	Airless Sprayer	600	(8)	
297	Tools	Pressure Washer	Pressure Washer	2000	(9)	
298	Tools	Pump	Domestic Water Pump	300	(4)	
299	Tools	Pump	Dual-speed pump	290	(5)	
300	Tools	Pump	Filter pump (1/2–1 hp)	1120	(5)	
301	Tools	Pump	Filter pump (2 hp)	1500	(5)	
302	Tools	Pump	Inflator Pump	50	(8)	
303	Tools	Pump	Irrigation/pond/waterfall circulation pump [1]	150	(2)	
304	Tools	Pump	Pump – Sump	1050	(7)	
305	Tools	Pump	Single-speed pump	540	(5)	
306	Tools	Pump	Submersible Water Pump	400	(4)	
307	Tools	Pump	Sump Pump (1/2 HP)	500	(6)	
308	Tools	Pump	Sump Pump 1/3 HP	800	(8)	
309	Tools	Pump	Sweep pump (3/4 hp)	560	(5)	

310	Tools	Pump	Variable-speed pump	120	(5)	
311	Tools	Pump	Water Pump (1/2 HP)	460	(6)	
312	Tools	Pump	Well Pump - 1/3 1HP	750	(1)	
313	Tools	Pump	Well Pump 1/2 HP	1000	(8)	
314	Tools	Router	Router	7	(3)	
315	Tools	Sander	3" belt sander	1000	(9)	
316	Tools	Sander	9" disc sander	1200	(4)	
317	Tools	Sander	Belt Sander	1200	(8)	
318	Tools	Sander	Disc Sander	1200	(9)	
319	Tools	Saw	Band Saw – 14"	1100	(3)	
320	Tools	Saw	Chain Saw - 12"	1100	(1)	
321	Tools	Saw	Circular Saw – 7-1/4"	900	(3)	
322	Tools	Saw	Circular Saw 8-1/4"	1400	(3)	
323	Tools	Saw	Miter Saw 10 inch	1800	(8)	
324	Tools	Saw	Reciprocating Saw	900	(8)	
325	Tools	Saw	Saw - Circular, 7 1/4 in.	1400	(7)	
326	Tools	Saw	Table Saw 10 inch	2000	(8)	
327	Tools	Trimmer	Hedge Trimmer	450	(3)	
328	Tools	Trimmer	Weed Eater	500	(3)	
329	Tools	Welder	Welder – Electric	9000	(7)	

Wattage Chart Data Sources:

(1) https://unboundsolar.com/solar-information/power-table
(2) http://hes-documentation.lbl.gov/calculation-methodology/calculation-of-energy-consumption/major-appliances/miscellaneous-equipment-energy-consumption/default-energy-consumption-of-mels
(3) https://unboundsolar.com/solar-information/power-table
(4) https://www.daftlogic.com/information-appliance-power-consumption.htm
(5) https://www.siliconvalleypower.com/residents/save-energy/appliance-energy-use-chart
(6) https://cornhusker-power.com/rebates/household-appliances/
(7) https://www.absolutegenerators.com/generator-wattage-chart
(8) https://www.batteryequivalents.com/boat-camping-rv-and-household-appliances-wattage-charts.html
(9) https://qlabe.com/heaters/power-consumption-chart-for-electrical-appliances
(10) https://www.nyseg.com/wps/wcm/connect/4e9ff288-f964-413b-a050-0f0535755915/13-0061+NR+Home+Energy-Use+Guide+Sheet.pdf?MOD=AJPERES&CACHEID=ROOTWORKSPACE-4e9ff288-f964-413b-a050-0f0535755915-mj8pTBk

Using a Kilowatt meter for measuring power consumption on individual appliances

A Kilowatt meter is a device that plugs into the wall, and then you can plug in your various appliances in for a true reading. Earlier I mentioned the Poniie PN2000 which I have used for my own measurements. Harbor Freight tools also offers a brand they call their "Kill A Watt" Electricity usage monitor. Both devices work the same – You plug the monitor into the wall, and then plug the appliance into the meter.

Harbor Freight Tools Kill-A-Watt Electricity Monitor Meter

By plugging in the meter and allowing it to track your appliance's power use for a couple of hours you will get an accurate measure of the appliance's power use. This is particularly important for appliances such as refrigerators or portable AC units where power use averages, we provided above can be considerably off from your actual appliance's need.

Both devices allow you to display the cumulative power through the meter since the power was applied, as well as then number of hours since power applied.

Notes about Running Watt vs. Surge Watt ratings

While shopping for a generator system – either a fuel-based generator (Gas, Propane, Diesel) or a Solar generator system, you will often see the generator model reference a number equivalent to the running or surge rating of the generator. The running rating of the generator refers to how much sustained power the generator can always put out running. This is the rating you need to focus on. The surge rating refers to how much power the generator can put out for a short period of time without damaging it. An example of a surge would be an Air Conditioner. When starting, and Air Conditioner pulls greater amounts of power than needed during sustained running. This is why in an RV some RV'ers can get away with a generator such as 1800-watt generator when their AC unit takes less than that to run but may have a starting surge power need of a few hundred watts more.

Very often fuel generator model numbers, such as my Predator 3500 generator, reference the surge rating and not the running rating whereas solar generator manufacturers model numbers will often do just the opposite – such as Maxoak's ACP200P which puts out 2000-running watts but can surge to 4000-watts.

MAKE SURE TO MATCH YOUR REQUIREMENTS TO BE BELOW THE RUNNING WATT RATING OF THE GENERATOR – NOT THE SURGE WATT RATING.

Pure Sinewave vs. Modified and Square Sinewave Outputs

Today most solar generators/power stations will have a pure sine wave inverter built in. This is good because these are devices you may more likely use indoors to power equipment such as laptops, modern televisions, or sensitive medical equipment. The quality of the rising and dropping of the power in the AC signal is clean, sometimes cleaner than from the power company.

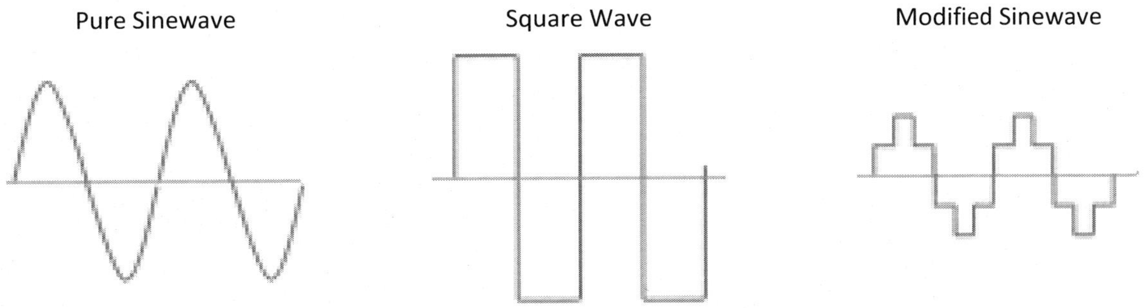

Table 8-2: Visual representations of three types of sinewaves – Pure which is the cleanest and our desired output, Square, and Modified.

With fuel generators, the lower priced open-frame and contractor types of generators will have modified sine wave output – harder on sensitive electronics and equipment with inductive loads.

IF WHAT YOU WILL BE POWERING IS SAFE TO RUN ON MODIFIED SINE WAVE POWER – THEN A LESS EXPENSIVE CONTRACTOR-STYLE GENERATOR WILL BE FINE WITH A MODIFIED SINE WAVE OUTPUT. BUT KNOW WHAT YOU WILL BE RUNNING WHEN PLANNING.

Either Closed frame or Open frame – will output a cleaner sine wave safe for your devices. When looking for a generator – Look for the terms "Inverter" or "Pure Sinewave". Some generators may also only say they have a "THD" or "Total Harmonic Distortion" of 5% or less. When looking at generators from some manufacturers they may not say "Pure Sine wave" but may refer to THD value. Generac for instance on many of their larger systems refer to the THD rating.

Below are Actual Sinewave quality output from multiple DC-to-AC Inverters I happen to own:

From Top left to right above: Bustek 300 Pure, Everstart 120w Modified, Everstart 750w Modified
From Bottom left to right above: Bustek 500w Pure Sinewave, Peak 400w Modified Sinewave

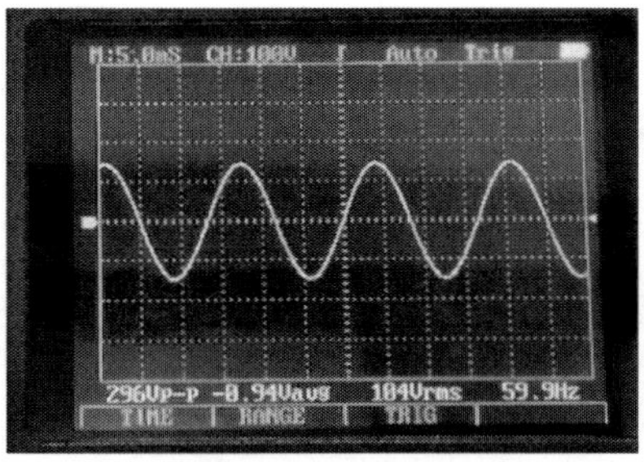

Bustek 300-watt Pure Sinewave

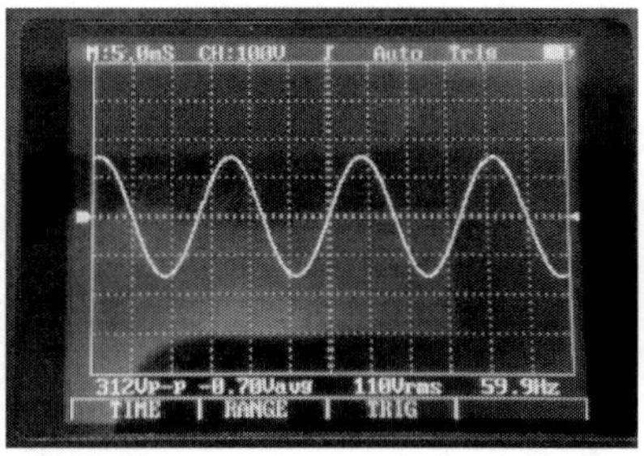

Bustek 500-watt Pure Sinewave

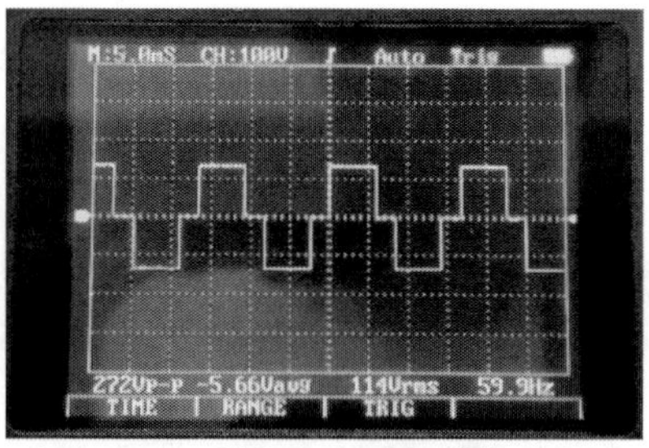

Everstart 120-watt Modified SW

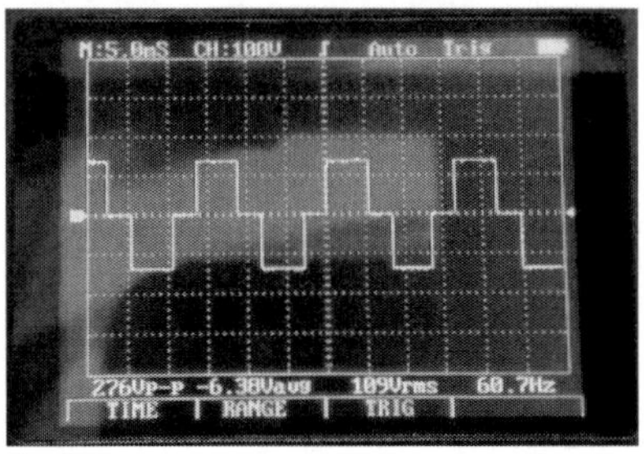

Everstart 750-Modified Sinewave

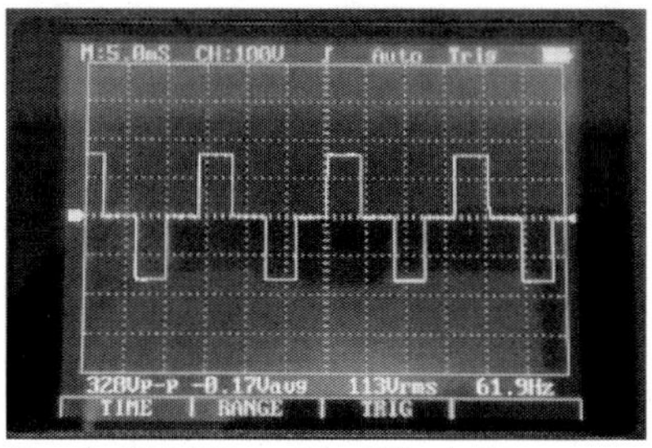

Peak 400-Watt Modified Sinewave

Lead Acid Batteries – State of Charge

When using Lead-Acid batteries (Flooded, Sealed or Gel) it is important to know when you have discharged your batteries below the 50% point. Use of a battery below 50% can damage the battery and greatly reduce the life of the battery. So – How can we tell the charge state of a Lead-Acid based battery?

By monitoring the voltage level of the battery, we can get an accurate reading that tells us the charge level of the battery. As shown in the chart below – if the battery voltage is at 12.7 volts or higher then the battery is considered to have a full charge. As it drains the voltage level will drop. Once it reaches about 12-volts (12.06 volts per the chart below) the battery is at 50% and further discharge should be avoided. Once the battery dips below 11.5 volts (between 11.31 and 11.58 in the chart below) you will begin to damage the battery which will affect the battery's ability to again recharge.

Lead Acid based batteries State of Charge

	State of Charge	Battery Voltage
Safe	100%	12.70
	90%	12.50
	80%	12.42
	70%	12.32
	60%	12.20
	50%	12.06
Caution	40%	11.90
Stop	30%	11.75
	20%	11.58
Damaging	10%	11.31
	0%	10.50

Lead-Acid Battery Stat-of-charge (SOC) Average Values

Inverter Fuel-based Generators

The Inverter Generator produces a clean AC output signal by way of an electronic inverter system built into the generator's output side. Because of the way the inverter works and the generator is designed, the generator can be set often in a quieter mode throttling down the engine allowing it to run more quietly, and also allowing it to burn less fuel than a non-inverter generator. When a generator can run on 25% to 50% of its running load capacity, it will be quieter stretching the time of use for the fuel in the tank.

Most smaller inverter generators are also enclosed in such a way that the case quiets the operation of the generator. This means the generator is more stealthy and harder to hear. In my RV I run an outside generator which is quiet enough that it can barely be heard sitting just ten feet away from the camper during the day while I am inside the camper. An open frame generator much loader. More inverter generators are becoming available however that are open frame which does reduce the cost but still maintains the quality of the power AC signal.

Backup Solution Reviews

Portable Generator System

Best for:

- Occasional power needs
- Least expensive (Can be under $1000)
- Consider Dual/Fuel (Gas & LP) for best fuel options during an event. Gas may be in short supply, Propane stores longer. Diesel is also a good alternative fuel for generators that will use it, however there are fewer small Diesel generators on the market.
- Portable – Easier to relocate if you have to relocate.
- Wide range of options – Contractor, Inverter, Multiple fuel types, wide price ranges.

Limitations:

- Dependent upon fuel to run.
- Requires maintenance – Additional oil for oil changes, spark plugs, filter cleaning.
- Shorter running times – Most cannot run beyond 12-hours at a time
- Cannot operate indoors – These require proper ventilation. Fumes are deadly.

Whole-Home Fuel-based Generators (Generac, Champion, Cummins, Kohler and others)

Best for:

- Whole-home power solutions
- More expensive but highly reliable – Cleaner burning fuels help reduce maintenance
- Not portable
- Requires permitting for installation
- Available to operate on multiple fuels – LP, Natural Gas, Diesel
- Longer running – Many models can run for Days

Limitations:

- Not portable – Cannot be taken with you if needed
- Natural gas is generally reliable but could be problematic in large scale power events or during deep freeze events such as the February 2021 Texas Freeze. Also – Natural gas still relies on a utility infrastructure to be operating properly.

Small Solar Systems – Portable Solar Generators/Power Stations

Best for:

- Operate well indoors – No fumes
- More portable than most portable generators
- Renewable energy – Repower from Solar panels
- Cheaper to operate long term (After initial investment)
- Some models support expansion batteries
- Large portable units can power RVs, Off-grid cabins
- These can provide power during rolling blackout periods without solar panels – Can be recharged from AC power when power returns (Saves cost of Panels)

Limitations:

- Less power output than a fuel generator
- Takes time to recharge
- Not as good for large power needs due to slower time to charge
- Expensive – Costs for medium to large between $1,000 and $3000 usually
- Solar panels need good sunlight for peak operation – Can be expensive, and performance dependent on weather

Large Solar Systems – Permanent Home Installations

Best for:

- Everyday reduction of electric costs
- Can provide whole-home power for off-grid with Battery supported design (Not Grid-tied only solution)

Limitations:

- Most Grid-Tied systems do not include batteries – These systems only work by feeding back to an active grid for energy cost reduction.
- Adding batteries of sufficient capacity for a whole-home solar system can be very expensive.
- Large solar solutions with sufficient battery capacity are extremely expensive running into costs exceeding tens of thousands of dollars
- NOT portable – Great for installed location but cannot be moved.

Ampacity values in wires and cable

The term "Ampacity" refers to the ability of a wire type to carry amperage or current. The number of Amps that a particular wire, including the wire jacket and insulation, varies depending upon the size of the wire as is shown in the chart on the following page. Pushing too much current through a wire can easily create a hazard by causing the wire to heat up to a level that the high current can trigger damage to the wire or fire. If wires are grouped or bundled this will affect how many amps or current can pass through the wire.

Copper Cable Gauge Size (AWG/mm^2)	Ampacity (A)
14 (2.08)	20
12 (3.31)	25
10 (5.26)	35
8 (8.36)	50
6 (13.3)	65
4 (21.1)	85
2 (33.6)	115
1 (42.4)	130
1/0 (53.5)	150
2/0 (67.4)	175
4/0 (107)	230

Fi Sample Ampacity values. Actual values are dependent on the specific cable you are using, makeup of the cable (Copper/Aluminum/Other), and the temperature of operation.

The Ampacity ratings of a wire will vary with temperature. The table is an example of literally over 3000 tables for cable and wire. The explanation of Ampacity if far beyond what we will be covering in this book, however our mention of it is to acknowledge that the ability of a particular wire to carry current will change based on size and temperature, and the hazards possible by not planning or adjusting properly.

National Electrical Code

Allowable Ampacities of Insulated Conductors Rated 0-2000 Volts

As Excerpted from the 2000 National Electrical Code

Ampacities of not more than three current-carrying conductors in Raceway, Cable or Earth.
Based on ambient Temperature of 30C (86F)

Size AWG or Kcmil	Copper Conductors Temperature Rating of Conductor			Aluminum Conductors Temperature Rating of Conductor			Size AWG or kcmil
	60 C	75 C	90 C	60 C	75 C	90 C	
14**	20	20	25	-	-	-	-
12**	25	25	30	20	20	25	12**
10*	30	35	40	25	30	35	10*
8	40	50	55	30	40	45	8
6	55	65	75	40	50	60	6
4	70	85*	95*	55	65	75	4
3	85	100*	110*	65	75	85	3
2	95	115*	130*	75	90*	100*	2
1	110	130*	150*	85	100*	115*	1
1/0	125	150*	170*	100	120*	135*	1/0
2/0	145	175*	195*	115	135*	150*	2/0
3/0	165	200*	225*	130	155*	175*	3/0
4/0	195	230*	260*	150	180*	205*	4/0
250	215	255*	290*	170	205*	230*	250
300	240	285	320	190	230*	255*	300
350	260	310*	350*	210	250*	280*	350
400	280	335*	380*	225	270	305	400
500	320	380*	430	260	310*	350*	500
600	355	420	475	285	340*	385*	600
700	385	460	520	310	375	420	700
750	400	475	535	320	385	435	750
800	410	490	555	330	395	450	800
900	435	520	585	355	425	480	900
1000	455	545	615	375	445	599	1000
1250	495	590	665	405	485	545	1250
1500	520	625	705	435	520	585	1500
1750	545	650	735	455	545	615	1750
2000	560	665	750	470	560	630	2000

National Electric Code Allowable Ampacities of Insulated Conductors
https://www.usawire-cable.com/pdf/nec-ampacities.pdf

Note: This is not a complete chart – This is only one of over 3000 data charts based on a variety of factors. Use for reference only.

New Technology to Watch

In recent years, we have had situation and types of events changing, and we have also been having very rapid leaps in the abilities of technologies. With such, we have new technology solutions that can be considered to provide solutions that are not as cut and dry as just a generator or a solar energy solution with panels and batteries.

Batteries are rapidly becoming more advanced, able to store more energy, lasting longer and costs are coming down. Just in the past 2-3 years advances in the Solar generator and solar power for home systems have greatly advanced. Below are some new types of systems that should be watched and considered.

Hybrid Inverter-Chargers

A hybrid inverter-charger is a relatively new component that is a single device that performs multiple functions with a full or partial solar solution. These devices will allow charging of a battery bank that is used for off-grid or down-grid power from either normal AC power when the power is up, or from a solar array also when the power is out. In fact, these units can be used without a solar panel array to keep a battery bank of lithium or lead-acid batteries charged up if the power problems you have are tied to rolling blackouts or short-length power interruptions. In California for instance, in dry season or during heavy loads rolling blackouts are often imposed that last only a few hours at a time. So without the expense or complications of a solar array, you can charge a battery bank for providing limited power to your home or certain appliances during the brownout, then when power restores your batteries are brought back up to full charge.

12: Resources and References

Below are a number of useful websites of interest on the topics of Generators, Solar energy, and emergency planning.

Websites

- Battery Hookup – Source for Lithium batteries, battery kits, build tools
  https://batteryhookup.com/
- Big Battery – Off-Grid Lithium Batteries
  https://bigbattery.com/
- Battleborne Batteries – High Quality LiFEP04 Batteries
  https://battlebornbatteries.com/
- Renogy Solar Products – Solar Products for Home, RV, Emergency Planning
  https://www.renogy.com/
- Victron Solar Products - Leader in Solar Products, Charge Controllers, Inverters
  https://www.victronenergy.com/
- ReLion Batteries
  https://relionbattery.com/products/recreational-vehicle-batteries
- Dakota Lithium Batteries
  https://dakotalithium.com/
- Mobile-Solar Power – Will Prowse – Reviews, DIY Solar
  https://www.mobile-solarpower.com/
- Generator Upgrade Supply
  https://generatorupgradesupply.com/
- Century Fuel Products – Gas to Propane Conversion Kits
  https://centuryfuelproducts.com/generators/conversion-kits
- Pinellas Power Products – Remote Controls, Fuel Extended runtime kits
  https://www.pinellaspowerproducts.com/
- How long can you run a generator for?
  https://www.generatorhero.com/how-long-can-you-run-a-generator-continuously/
- Natural Gas vs. Propane during the February 2021 Texas Freeze
  (https://www.forbes.com/sites/kensilverstein/2021/02/21/natural-gas-came-up-short-during-texas-freeze-but-propane-delivered/?sh=4bee484f7a3b

About the Author

Rodney Biddle is an author writing on topics of Small Power and Emergency power planning, Ham Radio, and RV Boondocking located in Florida. "Small Power Planning – Workshop" is his latest book project focused on producing a book for use in workshops looking at planning for backup power projects to use in the event of an event leading to loss of grid power. This book is designed to be a companion in workshop discussions of both fuel and solar solutions that will keep you up and running with electricity for periods without grid power if necessary.

Other books by Rodney include:

Small Power Planning: Emergency and Short-Term Power Planning

Are you ready to be without electricity for a few days? Or Weeks? Are you ready for possible Rolling blackouts? Small Power Planning is focused on helping the reader develop emergency and Short-term power solutions to help get through power loss. From unforeseen events from Hurricanes and Storms, Snowstorms, Rolling Blackouts, having an emergency plan for providing power for the conveniences to the necessities will help you have some level of protection. Electricity provides us comforts, necessities, and safety in our lives.

This book helps the reader to understand sizing and planning for various alternative power sources, focusing on Generator and Solar power solutions. We cover selecting a generator based on your needs, solar generators, and solar solutions for the home and portable power. Our book also covers RV and Camper solutions for off-grid boondocking which are also covered by most of our Small Power Solutions we cover in the book.

So, You just bought a BaoFeng Radio… Now What?

This is the book you NEED to go along with your BaoFeng Radio – We go into depth not just the controls of the radio, but also the How-To's and Real-Life Capabilities of the popular BaoFeng UV-5R/UV-82/BF-F8HP Radios. If you're a aspiring Ham operator we will help you to better use your radio to extend your range. If you're a Prepper – We cover what you need to know about the radio BEFORE you set it aside in your "Just-In-Case" drawer so if you need it you are really ready to go… This is the book that every New Ham with a Baofeng UV-5R/UV-82/BF-F8HP Radio should have, and the book that every prepper using these radios should keep with their radio for the emergency use reference with the radio.

So, You Just Got Your Tech License? Now What?

This book is written as a Next-Steps guide for new Ham Radio Technician License holders. Covering topics that go beyond what was in the test – Finding clubs, locating, and using Repeaters, Antenna limits, Radio Limits, getting more range from your radio, and touching on topics including WinLink and Echolink.

Ham and 2-Way Radio: Common Frequencies Reference Guide

Finally, a Frequency Reference guide to new Hams – This book provides a list of common frequencies for Ham operators as well as Frequency information for FRS/GMRS, MURS, Citizens Band (CB), Popular Shortwave, and Public Radio frequencies. More importantly it serves as a guide with references to popular tools used by new Hams for plugging in. This book helps you to skip some of the confusion and answer many newcomer questions to help you decide what you need to get into the right radio technologies to meet specific needs. We also cover comparisons between radio technologies – CB, FRS/GMRS and MURS for instance, and help you decide what is the right communications methods for your needs. Antennas and basic Antenna theory is covered with an explanation why these are so important to your successful radio operation.

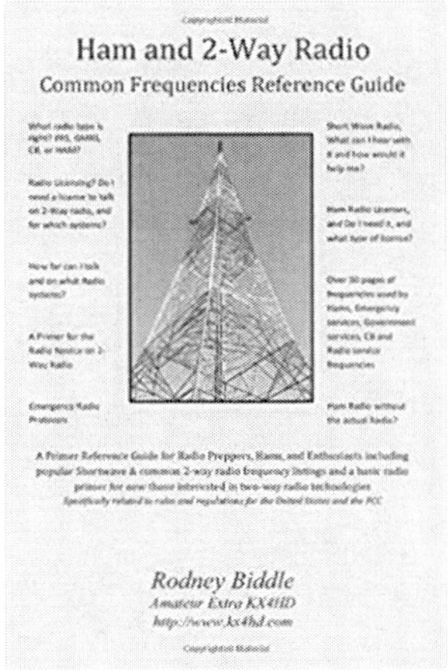

All these books can be found on Amazon.com

Made in United States
North Haven, CT
05 November 2026

21771721R00076